T0293534

THE PRIVACY MISSION

THE PRIVACY MISSION

MISSION

ACHIEVING ETHICAL DATA
FOR OUR LIVES ONLINE

ANNIE MACHON

WILEY

To my father, Nick Machon, who taught me about the need for freedom of expression and the courage to speak truth to power; and my mother, Michèle Mauger, who died two years ago, from whom I learned compassion and the need to fight for justice.

CONTENTS

ACKNOWLEDGEMENTS

I would like to thank the World Ethical Data Forum and Foundation for its support in writing this book, not least the founder and CEO, John Marshall, and our privacy expert, Sander Venema. An honourable mention should also go to my colleagues at Sam Adams Associates for Integrity in Intelligence, and also a vast array of hacktivists, journalists, politicians and technologists, whose company I have very much enjoyed over the years and from whom I have learned so much.

Last, but not least, I think both MI5 and David Shayler deserve a mention. Without those experiences, I should never have started down this path of discovery. . .

INTRODUCTION

The subject of data ethics and how the digital world impacts our privacy and basic rights has never been more urgent. This is no longer an academic or niche geek issue as it had been since the inception of the internet and the World Wide Web. Digital and data ethics are issues that affect all our lives, particularly as we have been forced to live increasingly online due to the Covid-19 pandemic.

We all have to start thinking about who controls the manufacture of and access to the hardware (including how it's manufactured and where the raw materials come from), who runs the software, who can spy on us, who can hack us and who can data farm us. These are issues that we, at the very least, need to be aware of in modern society. We need to ask what threats we need to protect ourselves from democratically, socially and personally. While there is certainly an element of individual responsibility, it is also essential to turn the lens onto business and governments. How can corporations protect us rather than prey on us, and how can so doing help their bottom line?

In this book, I aim to explore both the overarching concepts and principles about why this is important, and offer practical solutions for companies, policymakers and individuals to empower them to push back against these known threats, as well as to future-proof themselves going forward.

Technology is developing and expanding exponentially. The rate of change is unprecedented and it is only accelerating. What you will find in the coming chapters is information we all need to know and apply in the fast-evolving technologically driven world we now live in.

Untangling the World Wide Web

One of the challenges for anyone discussing our rights, privacy and data ethics in this digital age is unpicking the many threads that form the web of both threats and solutions in our digital world. As we move through the chapters, I will untangle some of the main threats we all face and explore solutions that can improve our privacy and safeguard both our data and our human rights.

I like to visualise this process as a starburst, with our rights, privacy and data ethics at the centre. A multitude of threats, choices and solutions shoot out of this starburst, which represents a utopian future where we are safe and free to live as we please, both on- and offline. These concepts, stories and ideas all interlink, too, creating the web that now supports so many aspects of our lives. Some of the threads within this web can be sticky, and others can lead to unexpected places.

To help you navigate the final section, where I present the best solutions available to us at present to reduce these threats, I have split the information as best I can to relate to individuals, businesses and governments. In each of these chapters, I explore the solutions that are most applicable at each level, although some of these invariably interweave. There is one common thread that connects all of these areas, as well as many of the solutions and threats I will talk about: education.

Education for an Empowered Future

The internet and online world as a whole is, in many ways, amazing. It has allowed us to live our lives like never before and gives us unprecedented access to other people, cultures and communities around the world. We have the power to share and exchange information, perspectives and knowledge in an historically revolutionary/novel way.

I use technology and appreciate all that it brings to my life. However, through my own experiences and journey of discovery, I know how damaging it can be to live in a world without privacy, one where you feel your basic human rights have been eroded. Rather than warning you off technology, I want to help educate you so that you can live your life, on- and offline, in a way that empowers you and protects your human rights, including your right to privacy.

While I will explore the multitudinous threats we face in our lives as a result of the shift to digital living, I also provide you with solutions that you can implement as individuals, in your businesses and even at government levels if you are involved here. One bastion of our privacy and rights in an online world is open-source software. This is one of the key countermeasures that has applications for us at every level of society.

Throughout this book, I will explain how open-source software (and hardware where it is available) can be applied to help guard against many of the threats that we face to our privacy and human rights. My aim with all the solutions I share is to give you a choice. I am not saying that you have to live your life online in a particular way. My hope is that by reading this book and better understanding the threats in your world you are able to carry out your own risk assessment and make choices about what levels you want and need to go to in order to protect your rights and privacy online.

For example, if you are a female journalist in Afghanistan or a human rights activist in Hong Kong, your threat assessment will probably be more stringent than that of a peace activist in Norway. However, your risk assessment is down to you and what I aim to do later in this book is provide options for your personal security.

Let's start by looking at why data ethics and online privacy are topics that should concern all of us, and why there is no such thing as "digital rights".

PART I
Privacy and Why We Should Care

There is no escaping it – we live in a digital world. In some ways, that is a wonderful thing. We are more connected to other people and cultures around the world. We have opportunities to learn from and work with people who are the best in their fields from the comfort of our homes. We can maintain and strengthen personal relationships, even with those who are geographically very distant from us.

While the internet presents many opportunities and technology has changed (and is continuing to change) the way we live our lives in positive ways, it is important that we do not dismiss and ignore some of the very real threats it can present, too. I am not here to tell you to bin your smartphone, delete all your social media accounts and disconnect yourself from the online communities you are part of. How much technology you introduce into your life is very much your choice.

As I have explained, I want to educate you about some of those threats and share practical solutions that you can implement if you see fit. This is about all of us feeling as though we have some measure of control over shaping our digital, as well as our physical, lives. You may never have considered your digital privacy before, or thought about how the rights you have online intertwine with your rights in the real world.

I want to help you see the whole picture, so that you are informed in your choices of both when and how you use technology in your life. In this first part of the book, I will explain why there is no such thing as digital rights, because these are simply our human rights, and share an insight into some events that have happened since the birth of the internet that have contributed to where we find ourselves today, in an increasingly digital society.

I also know from personal experience what can happen when technology is used against you to erode your privacy, which is why I am so passionate about our human rights in a digital world and ensuring we all have privacy, both off- and online. I share my story, as well as some of the things I learnt through my experiences, in Chapter 2.

Let's start by looking at our human rights in the context of our digital lives, what those rights are and why they are so important for each and every one of us.

CHAPTER 1
Our Human Rights Online

What does the term *digital rights* mean to you? Have you ever stopped to consider your rights in the online world and what those might (or might not!) be? Do you equate them with your human rights? If not, then now is the time to start, because there is no such thing as just digital rights. The lines between our physical and digital selves have blurred and that means our rights in a digital world are simply our human rights. You cannot separate the two; they are intrinsically linked. A quick history lesson into the origin of the Universal Declaration of Human Rights, which was put in place in 1948, will show you why.

The Universal Declaration of Human Rights was adopted by the United Nations (UN) General Assembly in the aftermath of World War II, when there was a strong focus on the social levers and mechanisms that had allowed Hitler to seize power and for Nazism to take root in Germany. The intention of this declaration was to put protections in place for both individuals and society, to mitigate against potential threats like this going into the future.

This was largely a mantle taken up by Western democracies, with other countries and cultures taking longer to go down the same path. In these Western democracies, there was very much a consensus that individual human rights were important to allow individuals and societies to push back against governments that might be sliding down a path towards totalitarian rule. The most important right within the declaration is, of course, the right to life.

However, in terms of our human rights in a digital world, the two that are most interesting and pertinent are the right to privacy and the right to freedom of expression. As a coda to that, the right to freedom of conscience is also important. Without those three rights, it becomes very difficult for society to reform, push back or move forward as necessary.

You only have to look back at movements for the abolition of slavery, universal suffrage or LGBTQ+ rights to see how important those rights are. If the people involved in those movements had been living under totalitarian, surveillance regimes, they would not have been effective. They would have been shut down and stopped – without freedom of expression, nobody else would have known what they were thinking. The point is, progress can be very easily stopped by surveillance and suppression. There are many examples where this is happening around the world even now, as I write this book. In 2020, Beijing imposed the National Security Law on Hong Kong, severely curtailing media and internet freedom in the territory and pushing Hong Kong towards an authoritarian system.[1] Under Saudi Arabia's male guardianship system, women are denied the freedom to make critical decisions about their lives and their freedoms are regularly curtailed.[2] Saudi Arabia also utilises its anti-cybercrime law to sanction those suspected of having extra-marital relationships as well as those in the LGBTQ+ community.[3] There has been a dramatic resurgence in digital authoritarianism in 2021, with Access Now and the KeepItOn coalition documenting 182 internet shutdowns across 34 countries. This is a sharp increase from the 159 shutdowns across 29 countries recorded in 2020. Ethiopia, Myanmar and India were among the countries to shut down the internet in a bid to quell the dissent brewing among their citizens.[4]

Free access to information and a place where people can express themselves freely is what enables people to mobilise and gives them the ability to connect and educate themselves about different issues, wherever they are based in the world. This, in turn, enables them to take a stance and campaign effectively to (hopefully) change the world, or the laws in their country.

However, to do this, journalists and dissidents need access to tools that enable them to communicate securely, because without these secure means of communication they are risking not only their liberty but also their lives.

These totalitarian regimes exist in our world today. They are not a theoretical construct or something conjured from the pages of a science fiction novel. Even if you do not live in one of these countries, it is not too much of a stretch to imagine you might visit a country where such a regime is in power and therefore be subjected to that regime's rules. Look at Nazanin Zaghari-Ratcliffe, who was detained by the authorities in Iran in 2016 and accused of being a spy. She has always denied the charges and maintained she was simply visiting her parents. Yet, she was imprisoned for six years and released in 2022 only after much political wrangling.[5]

Why Does This Matter Now?

Since the birth of the internet and its widespread adoption in the early 1990s, technology has developed rapidly. I remember those early days of the World Wide Web, when it seemed like some kind of utopia, where people could communicate, ingest and share information freely. You did not have to worry about online predators. It felt like a safe space. By the end of the 1990s the dot-com bubble had reached its peak and then burst, and this gave way to a new era of technological development, most notably around social media giants in the first half of the decade and then smartphone technology.

At the same time the technology was rapidly developing, the big technology and social media companies were just starting out. We saw the emergence of Google, Facebook and Twitter, as both corporate entities and the social media platforms we now all use and take for granted. The development and dissemination of smartphone technology made it very easy for us to press a button, download an app and live our lives through it. This has resulted in a very dangerous erosion of our sense of privacy.

I will explain precisely what these dangers are as we move through the book and share steps that you can take to protect against the risks you face. Awareness of these threats is the most important part, because if you do not know about them you will not even consider what you can do to protect yourself, or be able to assess whether you need that protection.

How do you feel about technology? How do you view the digital world? Your answers to these questions will likely be different than mine. They will also likely depend on when you were born, and whether, like me, you remember a world without the internet or whether you are part of a generation growing up as a digital native. If that is the case, you will have never known a world without the internet or this kind of technology. Anyone born since the early 2000s will have had access to technology and the online world from an incredibly young age. If you (or perhaps your children) fall into this generation, do you think about what that means for your privacy or your human rights? It is understandable if you have never questioned it like this before. For you, I imagine using the internet and technology is simply convenient, fun and an additional way of expressing yourself. Of course, it can be all of those things. Not all technology, or uses of technology, are bad; far from it. However, that is not to say it is all emojis, likes, shares, views and easy shopping experiences. There is a different, darker side to all of this that we need to be aware of, so that we can continue to use it in a positive way while feeling secure and protected.

Many of us (digital and non-digital natives included) do not think about the hardware or software through which we are carrying out a range of activities, whether it is downloading and watching films, socialising, having sex and relationships, or organising political activism. Many people also do not consider how much these activities say about them as a person, intellectually, emotionally, politically, sexually and in many other ways. Digital natives in particular do not think about these issues because they know no other way of being; this is just their normal.

In the past it has certainly been easy to think of our digital rights as separate from our human rights, but now that we live in an increasingly digital world, it is harder to pull apart the threads that separate the two. In fact, they have become intrinsically intertwined over the years, a process that was only accelerated by the Covid-19 pandemic.

Over the past 30 years or so, the internet and technology have developed exponentially, but until Covid-19 hit, there was still pretty much a choice in terms of how and when you used it.

You could choose which systems and apps you preferred, and how much of your life you lived online. However, when the Covid-19 pandemic struck, everything we did – be it health, finances, taxes, socialising or work – was forced online, particularly those of us living in Western countries. Using the internet was no longer a choice, it was a necessity. For many who found themselves working online, the systems, programs and apps they had to use were chosen for them by their employer.

This simply highlights why it is more important than ever before to consider our human rights in a digital context. Our online communications are an extension of ourselves and they need to be protected in the same way that our personal integrity in the physical world is. We require our human rights to be protected for every form of expression, both online and in the physical world, and if we are too complacent and just give up on those rights, we are putting ourselves in a very dangerous position, as I will explain in Part Two when I explore the Dark Triad of government agencies (spies), corporations and criminals.

We need to start thinking about our personal human rights and society's basic rights in terms of how we interact with this new technology. We also need to make sure we are not only looking at what has already happened, but also looking to the future to identify potential threats and try to protect against them to ensure there are laws in place to protect our human rights in this new, increasingly digital world. We also need to explore how we can gently nudge the cultural tanker in a different direction, as we did through universal suffrage, so that we, as individuals and a society, can take back some of our power and have much greater control over practices like endemic surveillance or data harvesting, rather than these being activities that need to be prevented by laws that could well be unenforceable.

The Privacy Versus Security Debate

For decades, intelligence agencies have trotted out the clichéd argument that if you want security, you have to be prepared to give up your privacy, but this simply is not true. In fact, if you want proper security, you need informed, educated and interacting citizens to act against potential security threats. In order for our citizenry to behave in this way, they need their privacy, be it in the real world or online.

It is only when citizens are informed, educated and able to interact with a certain degree of privacy that they can effect social cultural change and nudge the cultural tanker in a better direction. However, there is a balance to be struck and sadly it is one that all too often tips into Orwellian territory. The UK's Prevent programme is a good example of where the balance has tipped in the wrong direction.

Prevent was introduced during the height of the war on terror as a way for citizens to report concerns over radicalisation. Initially it focused solely on Islamic extremism, but over the years it has expanded to include considerably more activity on the far right, as well as any other potential terrorist threats. It is a draconian approach, because it requires people within communities to snitch on one another. It has also failed to prevent many terrorist attacks in the UK. An investigation by Channel 4 found that, of the 13 terror attacks in the UK between 2017 and 2022, seven were carried out by individuals known to the Prevent programme.[6]

The Prevent programme has also faced multiple accusations of systemic human rights violations[7] and has been found to inhibit the will of certain communities to speak up against extremist behaviour because they feel they are under undue levels of surveillance. This is clearly counterproductive and demonstrates why widespread surveillance is not the answer to many of the threats we face in society.

This is a "divide and conquer" approach, which is an alienating force. We saw this in East Germany, where the notional idea of trust was eroded within society because so many people were snitching on others that no-one felt they could communicate freely. This atomises society, rather than allowing it to be cohesive. To effectively guard against threats, we need to create a cohesive democratic community, where we can take action politically and potentially make a meaningful difference, because this makes us all more invested in that community and democracy.

However, we know that intelligence agencies and governments have used widespread surveillance on activists throughout the decades. Certainly it was a prevalent tactic used against the civil rights movement and people like Martin Luther King in the United States in the 1960s and 1970s. Similarly, MI5 dedicated a great deal of resources to surveillance on members of the Communist Party of Great Britain and a bewildering array of Trotskyists, as well as many other groups deemed to be subversive between the 1940s and 1993.[8] The difference is that back then it was very labour intensive to investigate and stop a very effective activist. Nowadays, it merely takes the flick of a switch to get a complete picture of someone's life and everything they are doing online.

As well as technology and the online world making it much easier for surveillance to be carried out, they make it much easier to disrupt an activist's activities. You can start planting fake news, or deep fake videos, to discredit someone (I will share stories of how these tactics have been used later in the book). You can set up a bot attack on social media and take someone down reputationally. It is very easy now for these predators with their own interests at heart, rather than society's, to restrict society's ability to protect itself and continue moving forward positively. I am not telling you any of this to scare you, simply to make you aware of the risks we all face as our lives become increasingly digital.

As we move through the book, I will share some simple steps you can take, whether as an individual, a corporation or a government, to help you protect your human rights online and make more informed choices about the technology you use in your daily life, whether that is a particular piece of hardware, a piece of software or an app.

Why Should I Worry About My Online Privacy If I Have Nothing to Hide?

Another argument that often gets thrown into the mix when people discuss human rights and privacy versus security is that if you have nothing to hide, then you have nothing to worry about. But now that many of us, certainly in the Western world, are largely living online, our digital selves are increasingly overlapping with our physical selves. It is a rather inelegant example, but would you want someone to be able to watch you while you were on the toilet? That would be considered a gross invasion of your privacy. Or do you want someone to have access to all of your financial records? Again, likely not something you would be happy about being accessible to other people.

But this is what the digital world offers, particularly in regions like Europe where we have ID cards that pull together various aspects of our lives and therefore allow a faceless bureaucrat to access pretty much any aspect of our lives, be it our financial records, health records or tax records. These are the kinds of things that most of us would prefer to keep private, particularly our health records.

As an example, in 2020 a private company that runs psychotherapy centres across Finland was hacked and the confidential treatment records of tens of thousands of patients were put up for sale online. The people whose records were stolen were not doing anything wrong, but nobody wants confidential details from therapy sessions to be available online for all to see. This is private information and should be kept private.

Now think about all the private information you readily make available in the digital world. I am sure you know that smart home speakers are listening all the time, effectively spying on you. There have been multiple stories from some of the big players (like Amazon) in the home smart speaker space where employees revealed they were encouraged to listen in to people's homes, to check that the technology was working.[9] But that is incredibly invasive.

There is also evidence of smart speakers being activated by mistake, listening in and recording when they are not supposed to. In one trial of various different smart speakers, researchers recorded one erroneous activation every five hours.[10] This begs the question, how much information are these companies collecting about you if they are recording your speech or conversations even when you have not activated them?

Similarly, the rise of video calls and long-distance relationships means that many more couples have explored online sex. They are not doing anything wrong and this should not be anything to be embarrassed about, but there are programs available, such as Optic Nerve that was run by British intelligence services, that allow images from video conferences to be captured, regardless of their nature. In fact, the data to come out of the Optic Nerve operation revealed that around 10 per cent of video conference calls were of a sexual nature. The people having consensual sexual relations with partners in this way were doing nothing wrong, but how would you feel knowing that someone had watched you at your most intimate?

These are just a few examples of areas of our lives that most of us would prefer to keep private, even though we are doing nothing wrong, but that have become increasingly difficult to keep private as our lives have become increasingly digital in nature.

Many Western intelligence agencies have been grappling with trying to find the right balance between security and privacy for decades. They have been asking how to ensure accountability, transparency and proportionality in terms of what spies and the police can access to protect the vulnerable while allowing greater freedom for the majority who are doing nothing wrong. It is not an easy question to answer. Part of the challenge has been the vast acceleration in both the availability and use of technology since the big tech giants took off about 20 years ago, how that has allowed a greater information grab and how it has eroded our sense of privacy.

There are many strands to this particular thread of discussion, from vulnerabilities in software that have been used and abused by intelligence agencies, as well as criminals, to how corporations harvest our data, using us as the "products" that deliver their profit. I used to describe this as a pincer movement between the state and corporations, where there is pressure being applied on two fronts by these two separate groups. However, as spy agencies are finding it increasingly difficult to keep their own stashes of cyber weapons secure, this has turned into a three-pronged attack, with criminals also exploiting these vulnerabilities and, therefore, us. This is what I call the Dark Triad, which I will come back to in much more detail in Part Two.

The Consequences of Digital Convenience

Over the years, we have steadily given up our privacy in the name of convenience. We have readily accepted a "forbidden fruit" and many of us were blissfully unaware of the consequences of the choices we were making. In fact, since the end of World War II, successive generations – from boomers to millennials to Gen Z – have become very complacent and expected their normal way of life to just continue. This is particularly true in the West. This complacency means many of us are not urgently worried about human rights and in many cases simply do not understand how fundamental these rights are to protecting our way of life.

This is a marked contrast to the generation coming through in the aftermath of World War II, when people were incredibly sensitised to the need to be able to push back against totalitarianism. This desire to build a better world led to initiatives like the creation of the National Health Service (NHS) in the UK, where health care was free at the point of access for all. However, since this time, a multitude of threats have been used as reasons to slowly strip away the rights previous generations fought for.

There was the war on drugs, which started in the United States but has spread to many other Western nations, and that has been used egregiously to strip away all sorts of rights and to give a raft of powers to law enforcement. Then came 9/11, when this activity accelerated further still. The "war on terror" has inflicted all manner of horror and chaos on large parts of the Middle East and central Asia, and led to a scenario where basic human rights are far from guaranteed for millions of people through no fault of their own. At the same time, those of us in the West are giving away our human rights, one cookie at a time. We trade them for the next shiny new gadget or toy, often without a second thought.

We now have a situation where there are generations of people growing up who do not know a world without this technology and who do not give a second thought to what they are giving up in the name of convenience. In the West, we are brought up to believe that the world is broadly safe, that we are protected by our governments and that corporations act with our best interests at heart. At the moment, this may well be true. But things can change; governments can shift. As I write this, we have already seen a huge upsurge in right-wing, nationalist political parties across Europe.

In 2022, Emmanuel Macron won the French presidential election against Marine Le Pen, and it was a closely-fought race. While Marine Le Pen may now be a busted political flush, her party will not go away. The issues she campaigned on will not go away. As I write this, France is preparing for its parliamentary elections, where Macron's party could lose its majority, leaving him a lame-duck president. In Hungary, meanwhile, Viktor Orbán and his right-wing party Fidesz won re-election in the country's 2022 general election, securing his fourth term as prime minister.

The right-wing Law and Justice (PiS) grouping has been Poland's governing party since 2015, while Belarusian president Alexander Lukashenko has maintained close ties with Russia and Vladimir Putin, despite a raft of Western sanctions following Russia's invasion of Ukraine. Even in countries such as Denmark, Norway and Germany, we have resurgence of nationalistic, populist parties, and far-right ideologies seem to be on an upsurge in many parts of Europe, usually because of the immigrant crisis. You can see a similar trend on the other side of the Atlantic, with Donald Trump's continued influence in American politics.

What if these kinds of parties become more commonplace and get into power? If we live our lives oblivious to the risks and are complacent about our human rights and our digital privacy, we could be in serious trouble if the government in our country shifts in this way.

We have to think forward to the potential threats that can come with the evolution of tech, the evolution of societal politics and the evolution of corporations that have access to a vast amount of our personal data.

The other challenge we are increasingly seeing is that governments are unable to rein in the big tech corporations. Quite simply, these corporations are bigger and have more money than most states around the world, which allows them to continue following their internal agendas without having to bow to government pressure. Governments are dependent on the infrastructure, hardware and software that these corporations provide, making it increasingly difficult for them to disentangle their interests from those of the corporations.

Carrying Out Your Own Risk Assessment

Like everything in life, working out what changes you may want to make to the way you live your life online will come down to how you assess the risks I have discussed in this chapter (and others that I outline as we move through this book) and how much you are perhaps prepared to sacrifice some of the convenience you take for granted in order to protect your human rights and privacy. In actual fact, as I explain later in this book, there are many alternatives available that are not particularly difficult to use but that can have a significant positive effect on your digital privacy.

I want to make you aware of some of the alternatives that are out there and I would like you to carry out your own risk assessment to understand what threats there are and which ones might affect you. For many people in the West, the risks are likely to be fairly minimal and therefore convenience may trump your other concerns, although for journalists, lawyers, whistleblowers and activists, privacy is already a growing concern. However, it is important to be mindful of the different places threats can come from, whether that is criminal hackers or big corporations, and what those various threats look like.

By contrast, the risk assessment for, say, a female journalist or academic in Afghanistan or an anti-war activist in Russia will be very different. If the journalist or activist wants to continue to write and disseminate information, they face the very real threat of being killed, in which case their risk assessment will lead to them being infinitely more cautious with the technology they use in their day-to-day lives.

Particularly for those of us living in the West, it is also important to consider whether any of the activities we are involved in now that are perfectly legal might cease to be legal in the future. For example, now you might be a peace activist but in a few years, the law may change and you will instead be labelled a domestic extremist. This will open the door for all the laws in that area to be used against you and, in that case, your online privacy (as well as offline) and what you share digitally, or already have shared, will become very important to you.

Follow the Leader

When governments are considering changing laws to give them greater powers to access our data, they have to be very careful. A case in point is the UK's Investigatory Powers Act 2016, which retrospectively legalised all the illegal snooping that the Government Communications Headquarters (GCHQ) had carried out since 9/11. The law itself is very broad in scope and allows the authorities to carry out bulk metadata, bulk computer and bulk dataset hacking. This gives them the ability to grab vast tranches of data or to hack into anyone's computer anywhere in the UK.

When this law was passed in the UK, Russia and China introduced new snooping laws in their own countries that mirrored those in the Investigatory Powers Act 2016. When pressed, they simply pointed to the UK and said (and I am paraphrasing!), "They did it first. They are a democracy and we are just following their lead". The UK gave much more draconian and totalitarian governments a very useful justification for enacting similar powers that arguably are not going to be used for the benefit of or to protect their citizens.

On the other side of the coin, businesses can boost their bottom lines by being trustworthy, authentic and taking steps to protect the privacy of their customers. Many businesses are becoming aware of this as an issue, and, as more individuals start to explore how they can protect their online lives, human rights and privacy, there will be a growing market for companies that sit on the right side of the data protection fence.

I will come back to what businesses can do in this respect later in the book. Right now, I would like to explore why our digital rights and human rights are one and the same. Understanding that digital rights cannot be separated from human rights is vital and highlights why this is such an important topic to discuss. I will also share my story about how I came to be such a passionate advocate and campaigner for privacy and human rights in a digital world.

CHAPTER 2
Gamekeeper Turned Poacher

Imagine what it would be like to know someone was watching your every move. You have a constant sense that you are being observed, whether you are walking down the street, watching TV in your living room, or being intimate with your partner. Your every move is tracked and monitored. Everything, down to your coffee order (and where you buy it from), is being recorded.

This sounds like a scenario out of a science fiction novel, but I promise that this happens to people right now. It has happened to me. I can tell you from personal experience what it feels like to live without privacy . . .

It was January 1991 when I arrived for my first day working at MI5. I remember feeling nervous and excited about starting this chapter of my life. Little did I know then that it would end so dramatically following my resignation in 1996.

My former partner and colleague David Shayler also worked at MI5 and he not only witnessed crimes, but also incompetence, cover-ups and injustices that he felt he should go public about. I also became increasingly disillusioned by what I saw through my work, from mistakes being made to lies being told to the government. The ethical framework at MI5 felt like it was in tatters, with no notion of account-ability or transparency. David and myself both raised our concerns internally, following the appropriate channels. We were both told the same thing: "Follow your orders and do not rock the boat".

The turning point came when David was briefed about an attempt to assassinate Colonel Gaddafi in Libya that was being funded by MI6, the UK's foreign intelligence service. During this briefing, he was told that the operation was going all the way to the Foreign Secretary for clearance, which is required under UK law for operations abroad. He felt uneasy about the operation, but his concerns were brushed aside.

In February 1996, David received various intelligence reports from different sources revealing there had indeed been an assassination attempt on Gaddafi. It had failed, and innocent people were killed in the process.

An explosion occurred under a car in a cavalcade in which Gaddafi was travelling. However, the bomb went off under the wrong car. Not only were innocent people killed in the explosion, but more were killed in the ensuing shoot-out with the security services. David also learned that the Foreign Secretary had not been asked for, or given, clear-ance for the operation, making it illegal under interna-tional law. David once again tried to go through official channels and was told to stay quiet. It was at this point that we started having conversations about leaving the service and blowing the whistle.

In the following months, David and I had many conversa-tions about our options, whether we could do anything else, how we would protect our friends and families if we did become whistleblowers (including those still working in the secret service) and how we could manage it for ourselves. There was a great deal of soul searching for both of us. It was an agonising decision.

David knew that by talking to the media about not only the Gaddafi plot, but many of the other illegal activities he had witnessed in his time at MI5, he would be breaking the Official Secrets Act, taking on the full force of the British Secret State and facing prison as a result. We both knew the consequences, but we decided we could not stand by any longer.

David made contact with a journalist and spent months building up a trusting relationship. During this time, he spoke very little about it. I was aware something was happening, but he purposefully kept me in the dark about the details.

Eventually, in the summer of 1997, the first story from David broke (although this was not about the failed assassination attempt on Gaddafi). David told me three days before the story was due to be published. We decided to preemptively go on the run around Europe to ensure that David could remain free to argue his case. I remember running around London using different phone boxes to book hotels and withdrawing cash from various ATMs so we could avoid using our cards abroad, as we knew that was one of the ways the authorities would be able to trace us.

This was the 1990s, so techniques for surveillance were still relatively analogue. The authorities had the same powers for surveillance as they do today, but back then it was more labour intensive and took longer to achieve the same results they could get in a matter of hours or minutes today. Despite this, we still had to be incredibly careful. We could not tell our families what we were doing or where we were going. Overnight we went from gamekeepers to poachers and our next month involved being chased all over Europe by MI5.

Initially, we flew to the Netherlands and spent our first week darting around various rural towns, just trying to stay one step ahead. In the days after the story broke, David started doing TV interviews. I remember being at a posh hotel in Amsterdam meeting journalists one day, and the next morning it was a mad dash to get out because MI5 had traced the journalists' phones and were closing in. We later found out that we were only two hours away from being caught. We were, understandably, very para-noid during this period, which, while useful for helping us evade capture, was also incredibly stressful. We travelled across Europe by train, moving from cheap hotel to cheap hotel, wearing disguises and using fake names. We never pre-booked anywhere and only paid cash. It was a very sur-real experience.

After one month of constantly watching our backs, I decided I would return to the UK and turn myself in. I was arrested and spent six hours being interviewed in a terrorist suite before being released without charge. This gave me the opportunity to return to our flat, which had been smashed up in a counterterrorism-style raid, as well as to speak to our traumatised families.

During my week in the UK, David spent time looking for a suitable place where we could hide out in France. Through very circuitous communication routes, he told me to get to a specific train station in central France and that he would meet me there. He had found us an old Victorian-era French farmhouse and this became our home for the next year. We were able to have contact with family and friends in the UK, but only if we called them, because at that time the spooks could not trace your number if you called in from abroad. It was a different story if someone in the UK called us, and we knew all of our friends' and families' phone lines were being monitored. In fact, we think this is how they discovered our location a few months into our stay in the farmhouse. However, by this time the media storm had died down, so they did not come for us straightaway.

We lived under the spectre of this constant threat, never sure if this would be the day someone found us. It was not only MI5 we were worried about. Both David and I had worked on investigations into the IRA and some of the early stories published in the press related to mistakes made in these investigations. At this time, the Good Friday Agreement had not been signed and there was a very real possibility we could be targeted by the IRA.

During these months, David worked with his journalistic contacts to break the story of the failed Gaddafi assassination attempt through the Mail on Sunday, which carried his original story, and BBC's Panorama.

In the summer of 1998, the Gaddafi story finally broke and we were thrown back into the spotlight.

We travelled to Paris to meet with a journalist about the story. While we were there, David was snatched off the streets by the French national intelligence service, known in those days as the DST. For three days, no one knew where he was. When we did eventually find out what had happened, we learned he was in prison awaiting extradition to the UK, because the British authorities had told the French that David was a traitor and that he was selling secrets to an enemy power. No one was allowed to see him at this time, not even his lawyer.

For nearly four months, we heard nothing. One day, following a high-profile court case, David was suddenly released by the French authorities. It transpired that when the French authorities had received his extradition paperwork from the UK they realised they had been lied to; David was not a traitor but a whistleblower. In France, this was deemed a political offence and the French authorities did not extradite for political offences.

This terrifying episode allowed us to move out of the farmhouse (which was basic, to put it politely) and live in Paris, much more openly than we had been before. During this time, I was free to travel between France and the UK, where I could speak to our lawyers and the media, and conduct political campaigning.

However, living more openly came with a price. The British security services knew where we were. We had friends and family visiting us, as well as journalists travelling to interview us. This meant we assumed that our flat in Paris had been bugged. We were, of course, already aware that all our communications, from phone calls to emails, were being monitored, but now we were under surveillance, 24/7, in our own home.

This made it impossible for us to talk freely and openly within our flat. If we wanted to have private conversations, we would have to go out for a walk. We did not visit the same brasserie more than once, in case that too became compromised. We knew that, even in our bedroom, we could not guarantee our privacy. As an individual, feeling as though you have no privacy is a very corrosive process. As a couple, this naturally has a damaging effect on your relationship too.

By August 2000, David felt he had done as much as he could from overseas and took the brave decision to return to the UK to face the music. He knew he would be arrested on arrival in the country and that he would go straight to prison for breaching the Official Secrets Act 1989, because there is no legal defence under this act. In November 2002, his case eventually went to trial, after which he had to spend a further six months in prison. This was a good outcome, given that he was facing six years.

By 2003, David had paid his debt to society and we were both free to travel. However, the challenge then became, how do you rebuild your life after such a period of intense stress? It was hard for both of us and, in 2006, we separated. I found myself asking how I could rebuild my life, what I had learned from that period and how I could move forward and help others.

This is really why this book exists today. The time I spent on the run and living under constant surveillance taught me how much I value privacy and also demonstrated how easily it can be eroded. One of the big lessons I took away from this period of my life was how much our online behaviour matters and how easy it is to fall victim to all kinds of predators, simply by failing to take basic steps to protect our privacy when we are using digital and electronic devices.

Going on the run from MI5 is not the only way to develop an understanding of how easily our privacy can be invaded in this digital age. In the wake of Edward Snowden's disclosures, a poll conducted by the Center for International Governance Innovation found that 43 per cent of respondents had changed their online behaviour as a result of what Snowden revealed, such as by being more careful about the sites they accessed and changing passwords more frequently.[1] This became known as the Snowden effect, although in the almost a decade since his revelations, this has died down.

WHISTLEBLOWER FILE: EDWARD SNOWDEN

In 2013, Edward Snowden went from being an unknown computer intelligence consultant in the United States to having his name splashed across all the major news outlets around the world. Why? Because, in a step of awe-inspiring bravery, he had taken highly classified information while working for the National Security Agency (NSA) in the United States and leaked it to the press.

What he revealed was a catalogue of human rights abuses by the NSA that occurred en masse and on a global scale and the global scale of the mass surveillance undertaken by the NSA in the name of protecting American security. We are talking about activities that go far beyond the dystopian horrors many of us are familiar with from reading George Orwell's *1984* (I share more details on the specific activities Snowden courageously unmasked in Chapter 4). The shockwaves from his revelations reverberated around the world and ignited a global debate about our privacy and how to protect it. One federal US judge went so far as to declare the NSA's activities unconstitutional.[2]

Despite many Western politicians reacting by shooting the messenger, some minor reforms to protect the rights of US citizens in their own country have been attempted, although none that come close to going far enough to remove the baleful gaze of the NSA and its vassals from the world's citizens.

Edward Snowden made a great personal sacrifice to bring these nefarious activities to the world's attention. Following his disclosures, he was granted political asylum in Ecuador, only to get trapped in Russia when the American government revoked his passport while he was on his circuitous journey to Ecuador. After spending weeks in hiding in Moscow airport, Snowden was granted asylum in Russia, where he lives to this day. He is still wanted by the US government, who accuse him of violating the Espionage Act. If he were to be extradited and found guilty, he would face up to 30–35 years in prison, despite the fact that his revelations were demonstrably of the gravest public interest.

As I started to rebuild my life in 2006, I became very out-spoken around issues relating to the "war on terror" and conducted media interviews about human rights abuses like torture and rendition. I travelled the world through my campaigning, speaking at events and meeting others who were campaigning for better human rights for us all in this ever-more-digital world, as well as for peace. One of the people I met through these events was former CIA senior analyst Ray McGovern. Now a peace activist, McGovern has also set up the Sam Adams Award for Integrity and Intelligence. I became involved in this organisation, which connected me to a global network of intelligence, government, military and diplomatic whistleblowers.

I have met NSA whistleblowers Thomas Drake and William Binney, as well as WikiLeaks founder Julian Assange. I have also been in contact with Edward Snowden through this network. Swimming in these tech-whistleblower circles opened my mind to the level of abuse happening around the world, which has become stratospherically worse in terms of capabilities since the 1990s, when I was on the run and under surveillance.

It was through these connections that I also got to know people within the "hacktivist" movement in Europe.

HACKTIVISTS

Computer hackers who do so for the greater good, rather than with any kind of criminal intent.

In 2007, I was thrown into the deep end of the German hacker scene and spent a lot of time in Berlin, which was a real hub for the hacktivist movement. I learnt a great deal about cutting-edge concerns and technologies during this period. After Snowden's disclosures in 2013, as you can imagine, paranoia in these circles was at its peak, particularly due to his revelations about the likes of the PRISM program.

When you arrived at a house party, you would be asked if you had a smartphone with you. If the answer was yes, the phone would get put in a biscuit tin that would be put in a fridge, creating a makeshift Faraday cage to prevent it from snooping on conversations. People I knew would also host parties where hackers would help you with activities like getting rid of Facebook from your devices, because simply deleting your account does not mean it is gone.

I found the hacker scene fascinating and it opened my eyes still further to cyber-security issues that many, including the large tech corporations, were unaware of.

I have been delivering talks to various hacker and hacktivist conferences and events ever since. What I find particularly interesting is that around 10 years after I started giving these talks, I received an increasing number of requests from established, corporate cyber-security companies asking me to talk to them about issues I had been discussing elsewhere for a decade. They are a long way behind the bleeding-edge research being carried out by hackers and hacktivists to this day.

The fact that it has taken nearly a decade for the cyber-security companies to catch up with what the hackers in Europe discovered and were protecting themselves against shows just how vulnerable many businesses are. Cyber-security companies that work with corporate clients have a duty to get closer to this bleeding-edge research from within the hacktivist scene. These companies need to recognise how far behind the curve they are and accept that they have a lot to learn from the hacker and hacktivist communities. All the information and discoveries from the hacktivist community is available via Creative Commons. It is there for the taking. However, the challenge is that many of these companies did not realise they had such huge blind spots until recently (some still fail to see their shortcomings!) and therefore they were not even looking for this information.

Since my time in the hacktivist scene, a whole raft of other nasties have come out of the woodwork. Among them are the revelations published by WikiLeaks about the huge cyber weapons cache held by the CIA. Essentially this is a huge array of vulnerabilities to software and hardware that the CIA was aware of, but that it chose not to warn corporations about.

The fate of the NSA cyber weapons cache was even worse. It fell into the hands of a criminal gang called the Shadow Brokers, who sold it off to the highest bidders. In this instance, the spies not only held back information that could have helped corporations better protect our privacy, they also allowed this information to fall into the wrong hands – you can hardly argue that they are protecting our national security and economic well-being in the face of such events. Instead they are making us as a society much more vulnerable to various kinds of attacks on our privacy and data.

The reason I have shared my personal experiences here is to help you understand how I came to be aware of the threat our privacy and human rights are under, from three fronts (the Dark Triad of spies, corporations and criminals) and why I am so passionate about raising awareness more broadly in society. Some cyber-security companies are beginning to wake up to the threats hackers have been warning about for a decade. Some national and even international institutions, like the British government, are trying to get to grips with a more ethical approach that will protect their citizens. The EU is certainly taking great strides forward in this area and it is even an issue that the UN is exploring. But the question remains, how effective can these organisations be in the face of huge, global media and internet corporations? When you factor in the spy agencies and criminal hackers as well, you can see the scale of the challenge facing us as a global society.

Individuals are beginning to wake up to the vulnerabilities in our software, but the problem is that this is just the top layer, because our hardware is also wide open to abuse. In fact, this was one of the Snowden disclosures. Did you know that, since 2008, all hardware (from computers and mobile phones to USB cables) has had backdoors built into it (and that this continues to be the case)? This means that even if you are extremely tight with your software, your hardware itself will still be vulnerable.[3]

We also know, thanks to Snowden, that since the late noughties all the big tech giants (starting with Microsoft and ending with Apple after Steve Jobs' death) were wittingly or unwittingly allowing backdoors to be built into their systems by the NSA and CIA. We also know, due to disclosures from earlier whistleblowers like Tom Drake and Bill Binney, that there has been a concerted effort by the NSA since the late 1990s to acquire, as Bill Binney says, "total mastery of the internet" (a concept I unpack later in the book).

This created a perfect storm in the United States, because the spy agencies realised they were significantly behind the curve just as Silicon Valley was starting to take off, and they made a concerted effort to catch up very quickly.

COUNTERMEASURE: PGP ENCRYPTION

Hackers and activists have long been looking for ways to help all of us, as individuals, lead more secure and private digital lives. In the 1990s, Phil Zimmerman developed Pretty Good Privacy (PGP) encryption, which is a pretty secure encryption system for email. However, in 1996 the US government tried to ban PGP encryption, arguing that it was a cyber weapon and could not be shared with the world because it was too powerful.

The US government finally dropped their case and PGP encryption was launched for the world to use. At this stage, it was incredibly difficult to use unless you were a hardcore geek. That said, I remember David managing to install and use PGP encryption in 1997 (something that still impresses my geeky friends even today!), when we were stuck in our French farmhouse with only a dialup internet connection – back then a feat that I would equate to scaling Mount Everest after just a few light strolls in the park to limber up.

PGP encryption is still unbroken and many people, myself included, use it to this day for secure email communications. While it is far from a guarantee of security, using it demonstrates that you are considering the privacy of your communication with others and sets a good example. The more of us use it, the more others will start to think more carefully about the wider issues surrounding the privacy of their communications online. You will find details of how to use it in Chapter 9.

What I have learnt from my experiences of living without privacy, and what drove me to follow the path I am on now, is that living with this lack of privacy and the degree of paranoia it inspires is not only harmful to your personal psyche, but also to the people around you. When you feel like your every move is being watched and your every conversation listened to, no one feels like they can talk freely to you. This is incredibly damaging to all your personal relationships.

Being a whistleblower can also be incredibly damaging for your own mental health. When you put your head above the parapet and speak out, you are likely looking at professional ruin. In many cases, financial hardship also follows and so do mental health problems. This is why finding a support network is so important. There are whistleblowers who have come through their incredibly difficult journeys and are thriving on the other side. You do not have to do this alone.

WHISTLEBLOWER FILE: JULIAN ASSANGE

While not a whistleblower himself, it is very difficult to have a conversation about whistleblowing without Julian Assange coming up at some point. He is a very polarising figure, but he is incredibly important within the whistleblowing community, not only as a publisher, but also as a protector of whistleblowers. What he has achieved is to establish a high-tech publishing organisation to protect whistleblowers – WikiLeaks.

(continued)

(*Continued*)

WikiLeaks was established in 2006 and is a high-tech and award-winning publisher of news leaks and classified media, all provided by anonymous sources. The organisation specialises in analysing and publishing large datasets of censored or otherwise restricted official material relating to war, spying and corruption. It has never been found to report anything factually incorrect in the time it has been operating. To this day, WikiLeaks has never given up a whistleblower. Julian Assange is its founder and director.

As I write this, Assange is still being detained without charge in the UK and his case has been referred to the Home Secretary (Priti Patel), who will ultimately decide whether to approve his extradition to the United States. Although new evidence has come to light and his legal team intend to make "serious submissions" to the Home Secretary against his extradition to the United States,[4] the chances of Assange escaping extradition are looking bleak. If he is extradited (as is looking increasingly likely at the time of this writing), he will face the wrath of a vengeful American establishment (politicians over the years have variously called for his assassination or kidnapping, and the CIA spied on him when he had asylum in the Ecuadorian embassy) that has become hell-bent on destroying his life and making an example of him in the process.

Assange's fate should concern us all. If he can be prosecuted for publishing information very much in the public interest, this means all the legacy media feeding off the hydrant of information WikiLeaks provides are also vulnerable.

Should he be extradited and face trial in the United States, his case will be closely watched. If he is prosecuted, it will surely be the death knell for the concept of a free media that is able to hold those in power to account.[5]

I met Assange for the first time in 2009, which was before WikiLeaks published the *Collateral Murder* video[6] that put Assange firmly in the sights of the American government. Even then, the level of paranoia he operated under was extreme. You can only imagine how much that has escalated in the years since. You can see that he is under a great deal of mental stress and it is no wonder that his mental health is disintegrating under that level of pressure. What Assange has and is experiencing is at the extreme end of the spectrum; David and I only experienced this briefly, but I would describe it as a form of extreme mental torture.

Meeting other whistleblowers and discovering the hacktivist scene helped me realise that there are steps I can take as an individual to improve my own privacy, as well as steps I can take to help others do the same. The situation is not hopeless. There are ways we can push back, both personally and as a society.

On a personal level, we can take responsibility for our privacy online, learn about the issues we are facing and make use of the technology available that can make our digital lives more secure (you will find a comprehensive list of countermeasures you can implement, what level of security they provide, and how you can use them, in Chapter 9). Use these as a signpost to start your own journey of discovery, deciding what level of protection you require based on your lifestyle.

We also need to be mindful of what comes pre-installed on our devices and therefore what you might want to uninstall on your phone, computer, tablet or other device. Open-source software is one of the simplest switches you can make and one that will significantly improve your online privacy and security. It is now possible to buy computers with open-source operating systems installed on them, but someone with the right level of technical ability can uninstall Microsoft Windows and replace that with an operating system of their choice, such as Linux Ubuntu.

Of course, the choice is entirely yours. You can choose to live your life out there in the digital world, with no additional protection, but I would like you to be aware that this is a choice you are making rather than one you are sleepwalking into.

On a societal level there are whole communities that are willing to help people develop their skills, if you are willing to learn. As an example, there is the global phenomenon of cryptoparties, where publicly minded geeks host an event that you can bring your laptop, phone or other device to and they will help you uninstall and/or install whatever software you want. It is free and a great way to improve your digital security while meeting new people.

If you see your government introducing draconian powers that will strip away your right to privacy online (and often offline as well) and you are concerned, do not stay silent. In the UK, you can lobby your MP or find a campaign group that you can get involved with if you are of a mind to go down the route of becoming a campaigner or activist. Even as an individual, there are things you can do that can and do make a difference, and I share more of these as we move through the book.

CHAPTER 3
How Did We Get Here?

To understand how we reached the point we are at today, we need to understand the origins of the technology we use daily and take for granted. The very first iteration of the internet was the Advanced Research Projects Agency Network (ARPANET), which was first developed in the 1960s as an American defence system that then spread into academia. This system allowed different universities and research groups to connect and share their work more easily.

It was not until 1989 that Tim Berners-Lee developed the World Wide Web, which was designed to allow people all over the world to share information much more easily. This was when the internet started to take root.

I can even remember the first time some very geeky friends of mine in Cambridge showed me the internet. Back then it was a wondrous thing that allowed you to communicate with people via computers and where you would find pages that you could navigate with the help of a mouse. If you are a digital native reading this now, you will probably think this sounds incredibly odd, but you have to understand that this was all completely new to us.

When I joined MI5 in the 1990s, it was fully paper-based. It was not until 1995 that the service adopted Microsoft as its communication platform, and during the entire period that David and I were on the run we only had one computer between us, which he used exclusively. This meant I fell out of touch with the technology and how it had evolved.

In 2005, I started using tech to further my campaign work and then, in 2007, I found myself thrown headfirst into Europe's thriving hacker scene. My knowledge of technology and its capabilities went from 0 to 60 in a matter of seconds! By this stage, the World Wide Web had existed for over a decade and we had already been through the first dot-com bubble, where a great deal of the technology had been commercialised.

In the late 1990s, the technology exploded. What are now the global tech giants, such as Amazon, Google and Facebook, all started out during this period. I can still remember that one of Google's early slogans was "Do no evil" (oh, the irony!). From there, everything shifted incredibly fast.

What I find particularly interesting is how the internet started out as an academic tool and then became commercialised in the mid- to late 1990s. At this time, when the first dot-com bubble grew and then burst, the focus was on what you could sell with the internet. It was not until the 2000s that we saw the seismic shift from corporations looking at how they could sell products to us via the internet, to us (and our data) becoming the products.

Our data is what these global behemoths of tech companies have monetised and it is how they continue to operate. Most people freely share their data and this allows for very effective, targeted advertising. These companies know where you are, they know what you like and, unless you are using tools like ad blockers and cookie crunchers (countermeasures I explain fully in Chapter 9), they can pinpoint and target you very effectively. You might think this is a good thing; you are shown more of what you are interested in and less of what you are not. In some respects, this can be very beneficial and it is certainly more convenient to go along with it. However, accepting cookies with a level of awareness about what data you are giving away in return is very different to just clicking "accept all" because you do not care!

COOKIE INGREDIENTS

First, it is important to recognise that cookies are a legitimate and necessary tool to enable modern web applications and the internet as we know it today to function. They are plain-text files that contain small amounts of data. Legitimate uses include for online shopping carts – a website needs to remember both the products you add to your basket and their quantities. Without some way of remembering and storing what you put in your cart, the site has no way of knowing what should be in your basket when you come to check out. In this context, cookies allow you to browse around a site, make new requests and add multiple items to your basket before hitting the "buy" button.

However, although cookies have legitimate and important uses, they are also used for nefarious purposes by many organisations. This is where expiry dates are important. All cookies have expiry dates, but the problem is that some of these can be set a long way into the future. As long as a

(continued)

41

(*Continued*)

cookie has not expired (or been manually removed), it will stay on your system along with the data it contains.

It is the persistent nature of cookies, and the fact that they can be shared across domains, that makes them problematic. This is what allows ad networks to track your browsing history across domains and thereby map your interests. Ad networks that can run across domains (like Double-Click, which was bought by Google in 2008) can access all the cookies stored on someone's browser. Social networks, such as Facebook and Twitter, use a similar system on their "share" buttons, allowing these organisations to track your browsing history even if you do not have an account with them.

It is this use of cookies that causes many of the privacy issues we see online today.

You and Your Data Are the Product

Making money from selling people's data is now a well-known business model, whether that data is contact information or browsing history. How do you think Facebook grew so exponentially and became so rich? By selling our data to ad companies. Any free service available online will probably be using our data as their product by selling it to third-party organisations. This means that each one of us who uses this technology has become the product and we are being data farmed. What I would like you to understand is that the big corporations treat us like battery hens.

All the data we churn out is being used to generate profit for other people; as individuals we do not see a penny of it.

The General Data Protection Regulation (GDPR) legislation introduced in the EU in 2016 was put in place in an attempt to rein in some of the most aggressive data-farming practices. While it is good in theory, it does not seem to be that effective in practice. As the United States has not followed suit, there is little to deter big American corporations from continuing how they always have when it comes to their data collection, storage and sharing methods.

If you look at a list of the world's biggest companies in 2022, you will see that it is dominated by tech firms. Apple, Microsoft, Alphabet (Google), Amazon, Tesla, Meta and Tencent are all featured at the time of this writing.[1] How do you think many of those companies made their money? Using data – our data. Data is the new oil.

The drive to commoditise our data has often been likened to the oil rush at the end of the nineteenth and beginning of the twentieth centuries, when many huge American corporate fortunes were established. Next there were the bankers in the early 2000s (and this continues to this day), who Occupy rightly highlighted as contributing to the growing wealth disparity in Western society.

The Occupy movement was a global group of activists who protested in 2011 about the bailout of banks "too big to fail" post the 2008 financial crash. Occupy was the first activist group to be spied on with panoptic tech in the United States, and also deemed to be domestic extremists/terrorists in the City of London.[2]

The tech giants are merely continuing this trend of using data to build their wealth and, thanks to their huge wealth advantage, they are able to manipulate the sector to ensure their continued dominance. As soon as a rival technology appears that could threaten their business model, they buy the firm out. Sadly, this has resulted in many startups predicating their success on being able to develop a product and sell it to one of the big tech firms for millions or even billions.

This has contributed significantly to creating what could be argued is the greatest wealth disparity we have experienced as a society since pre-revolutionary France in the eighteenth century.

Although it is referred to as our data, I want you to realise that it is much more than data. This is your life. We live so much of our lives online that your "data" covers every aspect of you, from your thoughts, relationships, political beliefs or activism to your financial and health records. All of this information is online and it is all accessible. There is a huge blurring of lines between our physical lives and online lives, which is what makes us so vulnerable.

Those of us who grew up in a world without the internet, and who very clearly remember that time, may have a greater awareness of some of these issues. But if you are part of the generation termed digital natives (broadly, anyone born from the 1990s onward), this has always been your reality and you may never have considered the underlying concepts surrounding your privacy and human rights that I explored in the opening chapter of this book.

In fact, these days I would argue that the only privacy we have as individuals is what goes on inside our heads. As you saw in

Chapter 2, if a government or security agency comes after you, your home, your technology – your entire life – will be bugged. They will see and hear everything that goes on in and around your life.

WHAT DOES TRULY SECURE COMMUNICATION LOOK LIKE?

The only way to securely communicate with another human being nowadays is to take a piece of porcelain, put a piece of paper on top of it and cover it with a towel. You then lift up the towel enough to write your message on the piece of paper, get the other person to put their head under the towel to read it and then burn the piece of paper, grind it into dust and flush it down the loo or throw it into the wind.

The reason this is secure is because no one can hear anything, no cameras can see what you are writing and there will be no imprint on the porcelain. David and I communicated like this when we were on the run. I know that Edward Snowden and Julian Assange have used this method as well.

This might sound overly dramatic but this is how far down the line we have come. We now live in a world where so much of our lives are online that the whole notion of our privacy has gone.

Social Media Has "Groomed" Society

Social media has played a significant role in society's reaching the point at which it is currently, by grooming society to think of this lack of privacy as acceptable. We are told that we should put ourselves "out there", we want to be "seen" and "authenticity" is in itself a big commodity on the internet. We have been conditioned to strive for a big social media following, because we have been taught that when this happens we get the big bucks for advertising.

Not only has social media taught us that the more popular you are, the more money you get, but we also know that positive interactions on social media give us a dopamine hit. The more you share, the more "likes" and "comments" you get, the more dopamine hits you get, the more you want to share. It is a vicious circle. Justin Rosenstein, the Facebook engineer responsible for creating the "like" button, has even described "likes" on social media as "bright dings of pseudo-pleasure".[3]

Social media gives us self-verification, allows us to amplify our personal voices and makes us feel as though we are quids in both emotionally and, in some cases, financially.

However innocent it seems and regardless of whether we enjoy using social media and putting our thoughts out there, or even whether we know we are being data farmed and data harvested by these corporations, the use of our data does not stop with them. We also have to consider how our data could be accessed by more sinister parties, whether that is criminal hackers or state-level actors. This brings us full circle to those arguments I discussed in Chapter 1, about how most people have nothing to hide. However, as I hope you have now realised, it is not about whether you have something to hide, but about whether you want the world to potentially see aspects of your life you would prefer to keep private, such as your health or financial records. The more we put out there, the more vulnerable we are.

Data farming is what the big tech corporations do. They collect our data, they use it and they abuse it. They farm it out to whichever organisations they see fit, be they insurers, advertisers or governments. The big challenge is that these corporations have created monopolies that make it very difficult to bring about change.

When smaller companies come up with cutting-edge ways of protecting people's data, their privacy and their rights to browse anonymously online, they are either crushed or bought out. Facebook has been using this model for years. The company bought out both WhatsApp and Instagram as soon as they became a threat. This is the challenge we are facing as a society, because these tech giants will ruthlessly protect their domain when it comes to ensuring they continue to have free and easy access to our data.

Open-Source Software Is the Antidote

It is not all bleak, because the open-source community believes that the knowledge it contains should be part of the common wealth and used for the greater good. This means there are people advocating for us to have access to open-source software and hardware to help break the monopolies the tech giants have constructed.

For example, as part of its Digitalisation Strategy, the Dutch government has launched an open-source toolbox to provide civil servants with useful information to help them switch to open-source software where it is appropriate to do so.[4] This has come about thanks to a clever and targeted campaign by the open-source community in the Netherlands; the Dutch government came to view open-source software to be more stable and better for national security, as well as better for developing a knowledge and skills base that could support the economy. It was a win-win decision for Dutch society.

One of the broader issues of public sector organisations continuing in this vein is that many do not have the budget to continually upgrade their software and hardware. Using very old legacy software that is not patched leaves them wide open to cyber-security breaches, such as the WannaCry ransomware attack that targeted the NHS in the UK in 2017 (which I discuss in much greater detail later in the book).

WOULD YOU LIKE A COOKIE?

We are constantly bombarded online by popups asking if we will "accept all" cookies or if we want to "modify" or "customise" the cookies we accept. If you click on the "modify" or "customise" button you will see just how many organisations will be given access to your data if you blindly "accept all". On many sites you will find a list of dozens of organisations that will be able to access your data if you simply agree to these cookies squatting on your computer.

You often will not see what specific data these organisations will be allowed to access, but the list is a telling reminder of just how much information you are giving away with a single click, and often without a second thought.

The only way to ensure those cookies do not remain on your computer after you have accessed a website is to use a cookie cruncher, which "crunches" all those cookies once you close that website. If you do not use a program like this – most people do not – these cookies continue to infect your device and you end up with layers and layers of them, many of which also cross-reference.

We Are Still Living in the "Wild West"

Back in the 1990s, when the dot-com bubble was at its peak, the internet and tech industry was referred to as "the Wild West" because of a lack of regulation in the sector. While many regulations have been introduced since the crash of the dot-com bubble, I believe that we still have a "Wild West" and that it is worse than it ever was in the 1990s.

The big global tech giants are operating outside of any jurisdiction. We have seen governments trying to rein them in, only to realise they could not get anywhere, which is why there is a greater focus on using anti-monopolistic tools to try to break them up, although I doubt the success this will have either.

The other big challenge is that very few politicians truly understand, or even want to understand, these issues. Even if you encounter a politician who seems to be aware of these issues, often their understanding is very superficial and they do not truly grasp quite how deep the threat is. The combination of all of these factors makes it very challenging to turn this cultural tanker around and get it to change course.

In 2005, Rop Gonggrijp, a Dutch hacker who has since become a friend of mine, gave a seminal talk at the 2005 Chaos

Communication Congress (CCC) event, in which he stated, "We lost the war", referring to the battle for a utopian internet where information is free for all to access. Rop is part of the early cypherpunk generation, many of whom are now in their 50s, who made good money during the first dot-com bubble but who have taken that money and used it for good, because they could see the balance of power shifting with the emergence of the social media sites in the mid- to late noughties.

You Have More Power Than You Think

This may all paint a rather bleak picture, but it does not have to be quite so pessimistic. The truth is that there are steps each of us can take, as individuals, to protect our data and therefore ourselves. The challenge is that most people are unaware there is even an issue, let alone the potential solutions available. There are a number of basic steps that you can take to protect yourself, which I will share as we go through this book. I have already talked about PGP email encryption and open-source software (and I cover this in much more detail later). There are plenty of other small and relatively simple steps you can take in your own life to better protect your privacy. Once you are aware that this responsibility lies with you, you can do something about it and take back control of your own life online.

New technology and software will continue to evolve; this is like a digital arms race and we have to take responsibility for our own digital safety and privacy, because we cannot rely on the corporations or governments to do it for us. Even if new software companies emerge that genuinely make security and encryption a key part of their platform, there is no guarantee that this will continue to be the case in the future. Software gets hacked.

Businesses get bought out by larger corporations. We have to keep one step ahead and being aware of these issues in the first place is an essential first step in the right direction.

What World Do You Want to Live In?

As I write this, there is increasing discussion about the creation of a digital world, where we can all live alternative lives, be the best versions of ourselves and interact with all these wonderful people. If this new digital universe were being created under a Creative Commons licence, was all open-source and truly for the common wealth, without any concerns about our data being farmed, commoditised, monetised, snooped on or hacked by criminals, I would be much more supportive and positive toward these developments.

The fact is, however, that this new digital universe is not being created under a Creative Commons licence. It is being created by the large corporations that own the infrastructure and hardware that will build this digital universe. Even taking our digital privacy into our own hands by using open-source software will not change the fact that the infrastructure and hardware itself has all been compromised.

The question to ask yourself is: Do you want to live your life online, in a world where you lose your intellectual property, where everything you use and interact with is copyrighted and where your data can be accessed and manipulated from the outside?

We have to be very careful when we consider these future scenarios and make sure that we are all asking: Who is creating these digital universes? Who owns them? Who controls them and has access to them? Whatever you upload to this digital universe will be your personal identity.

We also have to be aware of when joining such a digital universe ceases to become a personal choice. Just imagine for a moment that we have another global pandemic, like Covid-19, which requires lockdowns of entire societies. By this time, the metaverse is up and running. Now, instead of having Zoom conferences, your business has chosen to use the metaverse for communication and you therefore have to join this digital world and create an avatar. All of a sudden you are forced to have that presence, rather than choosing to have it.

Before the Covid-19 pandemic, you might have thought that kind of scenario was far-fetched, but we have already seen how quickly businesses can adapt to new, more digital ways of working, and how they expect their employees to do the same. Again, this comes back to the questions around who is controlling these online spaces (just look at the controversy surrounding Elon Musk's acquisition of Twitter!), how they are protecting (or not) our data and privacy and who can access them.

The arrival of the metaverse only makes these questions and arguments for protecting our privacy and rights in online and virtual spaces all the more urgent. We have to consider not only the possibilities this new technology offers, but also the very real risks it poses to us as individuals and to society as a whole. Who will be able to manipulate the metaverse to prey on others (we have already learned how it is being used by paedophiles to prey on children)? Who could tamper with the software? Would this enable these people to make your virtual self do something you would never do (another step on from the deep fakes we look at more closely in Chapter 7)? Who owns the copyright to your online identity?

Even if the metaverse is not owned solely by one corporation, we still have to ask which corporations own the hardware and the physical spaces where our information is stored and whether they can be hacked by the spooks or criminals. How can we trust the corporations providing the infrastructure for the metaverse to protect us from such threats? How can we trust our governments to protect us from these threats and how will law enforcement protect us from the criminals? These are all very legitimate concerns and questions that we should be asking.

When it comes to the metaverse, we also have to consider where it will be allowed. In many ways, what I have discussed here are very Western-focused issues, but is the metaverse going to be able to operate in China, Russia or Iran? Will those countries block people's access to it? Right now, we cannot possibly know how the metaverse will evolve or how it will affect the global community, because it is still in such a nascent stage.

The dystopian version of the metaverse involves big corporations controlling all the major platforms, where people's data is farmed, harvested and sold for tremendous profits, allowing all of us to be increasingly manipulated and exploited by governments, corporations and criminals. However, a utopian metaverse is not out of reach, provided there is sufficient education about open-source alternatives; there is hope that we can realise the potential of this new technology while protecting our privacy and data.

PART II
The Dark Triad

The Dark Triad is a psychiatric term that is applied to people with a James Bond–style character, one combining narcissism, Machiavellianism and psychopathy. My version of the Dark Triad does not relate directly to those personality traits, but it seems like a fitting way to describe what I consider to be a three-pronged attack on our freedoms on the internet.

For many years, I would talk about the pincer movement between state-level actors (the spies) and copyright enforcers (the global media corporations). On the one side, there were governments and spies who were eroding our freedoms in the name of security and developing their own caches of cyber weapons. On the other side were the media giants looking for ways in which they could enforce twentieth-century copyright legislation. This, ironically, has now come back to bite the corporations as governments in Australia and the EU are now trying to enforce fair payment of media-copyrighted material replicated on platforms such as Facebook. Then I realised that, in the twenty-first century, there is a third prong to this attack: the criminals.

As well as criminal hacking, there is another danger that comes from this faction, which is when the cyber weapon caches built up by spies fall into malign hands. These cyber weapons can then be unleashed on the world and used for anything but good.

So, when I talk about the threats we face online in the twenty-first century, I now talk about the Dark Triad of the spooks, the corporations and the criminals. In this part of the book, I will explore each in turn, as well as looking specifically at media control and cyber warfare, which are ever-growing threats in the modern world.

All of these threats interlink, creating a web of threats that we have to unpick and navigate, as individuals, within business, and at a national and supranational governmental level. It is essential that we are aware of the threats we face in an increasingly digital society, otherwise we have no chance of educating ourselves and protecting ourselves and our basic rights as technology becomes more advanced and interweaves with even more elements of our lives.

CHAPTER 4
Spooks

I am starting with the spooks because I believe it is necessary to understand what they do in the background in order to understand the full range of attacks and vulnerabilities we face. Ostensibly, the intelligence agencies are there, certainly in Western democracies, to protect our national security, our economic well-being and us as individuals. However, as a result of their activities, which they carry out in utter secrecy, they actually create more vulnerabilities than they stop.

In Chapter 2, I referred to a quote from Bill Binney, the former technical director of the NSA, about the agency's aim in the 1990s to achieve "total mastery of the internet". In their drive to do so, they gather up pretty much everything on the internet, but that does not help them do their jobs any more efficiently. In fact, I would argue that it hinders their efforts to protect us, the people, because they are dredging up everything and therefore all the nuggets of intelligence that could preemptively protect society are being drowned out in a tsunami of information.

The 1990s – When the World Changed

The 1990s was when the world really changed for the spies because that was when there was a real acceleration across all elements of society to get onto the internet. Up until that point, if a spy agency wanted to investigate someone it was quite a labour-intensive, difficult process. Spies had to carry out a real-world investigation, which meant physically bugging people's phones, intercepting their mail, following them around and generally trying to piece their lives together from the information they could gather through these activities.

As we all moved our lives online, this process became considerably easier, to the point where it is almost as simple as the spies flicking a switch to suddenly have access to a wealth of data about a given individual. Social media has made it simpler still for the spooks, because all of a sudden we were all offering up our information, for free, simply because it is fun.

Now, the spies can very easily find out everything about us, from our relationships to our political beliefs and our work. It is not only Facebook, of course; all social media platforms provide the same sort of easy access to our information for intelligence-gathering organisations (and not just the official ones!).

Initially, certainly in the UK based on my own experiences as a spy, the intelligence agencies were slow to catch on and definitely had some catching up to do throughout the 1990s. On the other side of the Atlantic, however, it was a slightly different story.

Total Mastery of the Internet

In the United States during the 1990s, Silicon Valley was just taking off. We had the rise (and then fall) of the first dot-com bubble and the NSA realised that the internet and technology presented a huge opportunity. It was in this context that Bill Binney said that the NSA's aim at the time was "total mastery of the internet".

This is what the Americans set their sights on, and the NSA pressured the Global Communication Headquarters (GCHQ), who are essentially the UK equivalent of the NSA, to follow suit. The NSA also provided substantial funding to GCHQ during this period.

In the early 1990s, the Echelon programme emerged, which is now called the Five Eyes. Essentially it is collusion between anglophone spy agencies around the world, although most notably between the UK's GCHQ and the United States' NSA, which means they share almost everything. The way it works is simple. For many decades, it was illegal for GCHQ to spy on British citizens and it was illegal for the NSA to spy on American citizens. However, they could spy on each other's citizens and then share their information. Although it started as a relationship between the UK and the United States, the programme has been expanded and now also involves Canada, Australia and New Zealand (making up the Five Eyes).

What this means is that all five of these countries can snoop on each other's citizens and then share their information, thereby getting around any notion of democratic oversight within the countries themselves. Of course, in terms of the capability the spooks have for snooping, this activity has exponentially grown along with the technology.

WHISTLEBLOWER FILE: WILLIAM (BILL) BINNEY

Bill Binney joined the NSA in 1970 and spent the next three decades working for the agency, eventually becoming technical director, a post he held until his retirement from the

(continued)

(Continued)

NSA in 2001. That is not to say his time at the NSA was all smooth sailing.

In 2000, Binney, along with his former NSA colleague J. Kirk Wiebe, turned whistleblower. At this stage, they were still working for the NSA and approached Congress over their concerns about national security and the mismanagement of important technology programs. They believed that a program they had developed for the NSA (ThinThread) could potentially have detected and prevented the 9/11 attacks. However, their program was shelved in favour of the much more costly Trailblazer program, which failed to work and was ultimately scrapped, costing US taxpayers billions.

After both were demoted within the NSA for seeking Congressional oversight, they retired and, in 2002, made their concerns about the NSA's practices public.

ThinThread and Trailblazer

The ThinThread program was developed by Binney and his team in the late 1990s/early 2000s. It was designed to hoover up whole tranches of online data but, crucially, it filtered out data from US citizens within the United States, thereby protecting their constitutional rights. Its price tag was about $1.2 million. However, ThinThread was never used and instead a much more expensive program, Trailblazer, was selected. This program was created by a private company and cost the US government somewhere in the region of $1.4 billion.

However, unlike ThinThread, Trailblazer did not filter out the data of American citizens in the United States, making it unconstitutional under US law. What is more, it never even got properly off the ground, providing no benefit to the US intelligence services despite its hefty price tag.

We also know that in the aftermath of 9/11, when the war on terror started, the process of intercepting digital communications and data accelerated. Everything was done in the name of protecting nations against terrorists. It was not until 2013, when Edward Snowden made his first revelations, that the world suddenly realised the sheer scale of what the NSA had been doing in their pursuit of "total mastery of the internet".

It is not only in the United States where our freedom on the internet is being attacked. In 2016, the UK introduced the Investigatory Powers Act, which I mentioned in Chapter 1. This allowed GCHQ to legally carry out bulk dataset hacking, bulk metadata hacking and bulk computer hacking. However, it was applied retrospectively, essentially legalising all of these previously illegal activities all the way back to 2001 and the start of the war on terror.

What all of this shows is the sheer scope and scale of what spy agencies in the Western world can do, in that they can intercept pretty much anything and they can do so globally. I (and others) describe this as the global surveillance panopticon. The key question is how this data is managed, how it is put to use and how much of this data is actually accessed as opposed to just being stored.

The Snowden Disclosures

I introduced you to Edward Snowden in Chapter 2 (if you were not already familiar with him); what he disclosed about the scale of snooping on the internet sent shockwaves around the world. The following are just some of the most pertinent disclosures he made as a whistleblower.

PRISM

One of the first disclosures Snowden made was about a program called PRISM. This program gave the NSA, and by extension GCHQ, backdoor access to all of the technologies produced by the global tech giants. We are talking about the likes of Microsoft, Twitter, Google, YouTube and Apple.

What this meant was that, wittingly or unwittingly, these global tech giants were allowing the spies to snoop on all the information passing through these corporations. In terms of the meshing of the corporate and the state, this was an eye opener for many people.

Operation TEMPORA

Another of Snowden's key disclosures was about Operation TEM-PORA, which was another collaboration between the NSA and GCHQ. What Snowden released showed that GCHQ had agreed to hack all the fibre optic cables between North America and Northern Europe, for a sum of about $100 million from the NSA.

Essentially this meant GCHQ were prostituting themselves to the Americans and gathering petabytes of information travelling between the two continents. At the time that this operation was agreed, the so-called democratic oversight came from the Foreign Secretary in the UK, who had to sign a warrant to allow this information to be gathered. In doing so, he was giving GCHQ permission to snoop on not only the whole of the UK, but also the whole of North America and much of the rest of Europe. Following Snowden's disclosures in 2013, then-Foreign Secretary William Hague defended the UK's position. It is an anti-democratic step.

The Hacking of Belgacom

Another disclosure from Snowden was that GCHQ had hacked into Belgacom (now Proximus), which is Belgium's biggest telecommunications operator. It is used not only by people across the country, but also by the European Commission and all the other EU organisations.

Understandably, that caused outrage because these countries are the UK's allies; they are not the enemy in any sense.

XKeyscore

The final Snowden disclosure I will mention at this juncture is about the program XKeyscore, which also allows the NSA to access a huge amount of information. What made this disclosure particularly important was that it revealed that Germany's foreign intelligence agency (the BND) had worked with the Americans to develop it and had been using it themselves.

It later came out that over 100 German politicians, including then-Chancellor Angela Merkel, had had their private phones hacked using this program. This just goes to show the entirely indiscriminate nature of these powers – and there are many more examples, both from Snowden and other whistleblowers, that show just how broadly and indiscriminately security agencies spy on populations, both in their own countries and overseas.

Knowledge Is Power

If you are wondering *why* national intelligence agencies from around the world would engage in such activities, the answer is simple: knowledge is power. This has always been the ethos of any intelligence agency, not just in the West but across the world. There is very much an attitude that if we can get the information, we should grab it, because it might be of use in the future. We can certainly see that happening in China, with their mass surveillance programmes and social credit system. Russia has long been excoriated for just gathering everything they can through cyberattacks and hacks.

As I mentioned in the first part of the book, you might not be bothered that your data is being collected and stored. You are not doing anything wrong, so what does it matter? But here is the thing: You might not be doing anything wrong now, but what if the law changes? What if a government with more authoritarian tendencies is elected?

Imagine you are anti-war, so you go to a peaceful demonstration. You are just an activist who is exerting their right to freedom of expression. However, these demonstrations are always filmed, be it by police, CCTV or videos posted on social media. Facial recognition technology, as unreliable as it is at the time of this writing, exists. Our phone tracks our location and therefore our GPS data is also stored, proving we were at that demonstration. Right now, everything you have done is completely legal. You have nothing to worry about.

But what if a social emergency occurs, such as a pandemic, an economic meltdown, or a war, and the government starts to crack down on protests; new laws are brought in. Suddenly, rather than being seen as an activist who attended a peaceful protest, you are considered to be a dissident or even an extremist. All the proof you attended that demonstration has been stored and, now that the law has changed, you could be held to account or even prosecuted in the future.

Perhaps you are a business owner with certain social values and views that, while perfectly acceptable now, come under fire in the future. The ease with which this information can be gathered and used against you is staggering.

Societally, these are issues we all need to consider, because of the sheer scope and power of what is held on us now and how it could all be used against us in the future if things spiral in the wrong way. In fact, we have seen this kind of thing happen before, and we are on a path towards it happening again.

BLACKLISTING THE ACTIVISTS

In the 1970s, there was a series of mass strikes in Liverpool. At the time, those who were involved and considered to be extreme trade unionists and potentially communist sympathisers were investigated by MI5 and Special Branch, resulting in some of them going to prison. When these men came out of prison, they found they had been blacklisted and could no longer get jobs in their industries.

One of these men was Ricky Tomlinson, now an actor. He was blacklisted and could no longer work in his trade, which is why he went into acting. Even back then, when

(*continued*)

(*Continued*)

everything was paper based, he and fellow activists were blacklisted very effectively, forcing them into other careers.

Now imagine scaling that up, using the data that is being collected on all of us about our online lives, and you can start to see the sheer scale of the potential threat to all of us. In some cases, this has already happened. Look at the Occupy Movement in 2011, where many people joined protests against the financial sector and global wealth inequality in London. Through police surveillance many of those who attended the protest were identified and they now all have criminal records. Some of them were even deemed domestic extremists rather than activists.

In fact, a letter from the City of London Police that was sent to the banks in the area even said that the police viewed the people involved with the protest as "domestic extremists/terrorists".[1] That just goes to show how easy it is for a change of perspective, demonstrated through that change of language, to happen, and makes the concept of mass data collection about all citizens much more sinister.

We already know of a catalogue of abuses against activists, from the British police spending years spying on innocent protest groups and laws that allow activists to be preemptively arrested in the UK, to collusion between Western governments and spy agencies that facilitate the kidnaping, torture and assassination of alleged terrorist suspects (who could just as easily be activists).[2]

It is not only the future leaning of our governments that we need to worry about, though. In the West, we tend to think that the Western agencies are here to protect us and only do good. What they have done (particularly through the Five Eyes program) is create a global network, this global panopticon, that allows them to spy on anyone, anywhere. However, what concerns me more (and what should concern all of us) is that all of these agencies struggle to protect their own tools and their own data from the third prong of the Dark Triad, the criminals.

Even if the Western intelligence agencies only use our data to protect us (and I am not saying this is the case!), just imagine what might happen to it if it were to fall into criminal hands. As I mentioned in Chapter 2, this has already happened. The cyber weapons and tools hoarded by the likes of the NSA and CIA have fallen into criminal hands and once they are out "in the wild" these weapons mutate. It is terrifying that even the intelligence agencies are unable to keep their information secret in this online "Wild West".

Hacking the Spooks

In 2016, a group called the Shadow Brokers announced that they had hacked the NSA and had access to some of the agency's cache of cyberattack weapons, as well as zero-day hacks and vulnerabilities to the software and/or hardware of mainstream corporations. The Shadow Brokers proceeded to put some of this information up for sale, as well as publishing some of the remainder, which ended up in criminal hands.

What came as a shock to many was that the NSA had been aware of the vulnerabilities, or backdoors, in the software or hardware from companies like Microsoft, but that it had not shared that information with the companies. Essentially they left the corporations and their customers open to potential cyber-security breaches as a result.

One example of how these vulnerabilities were exploited came with the WannaCry ransomware attack in 2017, which affected many organisations around the world, including the NHS. Hackers were able to exploit a vulnerability in Microsoft Windows to introduce ransomware to whole organisations, encrypting important files and telling users that they would only regain access once they had paid a ransom in Bitcoin. One-third of NHS hospital trusts in England and Scotland were affected and it is estimated that the incident cost the NHS £92 million.[3]

Microsoft had issued a patch for this vulnerability nearly two months before the WannaCry attack went global. However, many organisations fail to regularly update their operating systems, meaning that they remained vulnerable to this particular backdoor.

In 2017, there was another example of an intelligence agency's cache of cyber weapons being leaked. This time it was the CIA and the disclosure came via WikiLeaks, under the codename Vault 7.

Those responsible for these events are ultra-secretive organisations that are storing up a raft of highly sensitive information, which could be of use to corporations or to individual citizens to help them protect themselves, as well as very aggressive viruses that can be used as cyberattack weapons. Yet despite the huge damage these vulnerabilities and cyber weapons can do, they cannot even protect their own caches. That begs the question, whom can we trust with internet vulnerabilities and information? If the CIA and NSA cannot keep their cyber weapon caches safe, who can?

The point is that, by holding these things back and then allowing them to go out into the wild, Western intelligence agencies, either wittingly or unwittingly, are not doing their job, which is to protect the economic well-being and national security of their citizens.

Cyber Weapons: A Pandora's Box

While the fact that the likes of the CIA and NSA kept vulnerabilities in widely used software to themselves is shocking, what is equally, if not more, concerning is the development of cyber weaponry.

We all know that Russia and China are engaged in developing cyber weapons, but intelligence agencies in the West are doing just as much work in this area. For example, in the UK they have now established a cyber attack force, in addition to the cyber defence force that sits within GCHQ. There is one story that is now more than a decade old, but that illustrates just how dangerous cyber weapons can be: the Stuxnet virus.

Stuxnet was jointly developed by the American and Israeli intelligence agencies; its sole purpose was to contaminate the software that was being used to run Iran's nuclear power facilities. Stuxnet targeted the centrifuges that are used to produce the enriched uranium required for nuclear power (as well as for nuclear weapons) and caused them to spin too quickly or for too long, damaging the equipment in the process.

Stuxnet was first discovered in 2010, but had likely been in development for years before. Since it was unleashed on Iran's nuclear facilities, it has been unleashed into the world, where it has mutated, much like a natural biological virus would mutate. This is the problem with computer viruses; although Stuxnet was developed for a very specific purpose and target, once its code was out there, it could be mutated and used for different purposes.

It can be very easy to think of states like Russia and China developing cyber weapons, while ignoring what goes on here in the West. However, as the Stuxnet example shows, the West is a leader in this field – and that is incredibly frightening.

Cyber Threats: What You Need to Know

The following threats are particularly pertinent for those working in corporate arenas, because you need to be aware of them to guard against them.

Spear phishing: Fundamentally, spear phishing attacks all rely on the same vulnerability: the idiot at the keyboard (you or I). However, gone are the days when you would get an email from a Nigerian prince telling you they would share their wealth and all you had to do was "click here". Spear phishing attacks are much more sophisticated than they used to be. For example, an employee might be duped into clicking on a link because they get an email that looks like it has come from their CEO. Much like with the Nigerian prince, this then downloads a virus. If the computer you are on is connected to a network, the virus can then infect the whole corporation, not just that one machine.

Zero day exploits: These are the vulnerabilities in the software of major tech corporations that were identified by spy agencies, but not shared with the corporations in question (the ones uncovered in the NSA and CIA cyber weapon caches, as I explained earlier). Therefore, the tech corporations did not provide patches for these vulnerabilities to their users because they did not know of their existence. As a result of the spy agencies keeping these vulnerabilities to themselves, presumably as potential assets they could use in the future, all the users of that software were put at risk of a hack. In some instances, like the WannaCry ransomware attack, there were serious consequences.

The Blurred Line Between State and Corporate

There is also an increasingly blurred line between state intelligence agencies and the corporate world, with any number of examples of "stooges" who act on behalf of a particular state; using them is a way of making the agencies one step removed from the hacking and cyber infiltration, while still getting the outcome they want.

Consider, for example, Cozy Bear, a group of Russian hackers who are believed to be associated with the Russian spy agencies. In the UK, our equivalent would be a military intelligence organisation called 77 Battalions. Regardless of who they are acting for or against, their role is all the same: to inject poison into the online world of adversary states. For those in the West, that would likely mean targeting the likes of Russia, China, Iran and North Korea. All of those states' intelligence agencies have ties with their own private organisations doing exactly the same in the West.

There is an increasing overlap between the corporate and the state, which can make it hard to separate one from the other at times. Of course, this only applies in larger countries that have an extensive, advanced intelligence structure. There are many countries around the world that do not have access to this level of resources and that are, in fact, preyed upon by these resources coming out of bigger, more powerful countries.

One of the most high-profile examples of how the state and corporate worlds have bled into each other comes from Israel's NSO Group, and the development of Pegasus, a particularly insidious program. Pegasus was developed privately as a spyware program designed to target mobile phones. In its early iterations, you had to click on a link for your phone to be infected with the Pegasus spyware. Once it was installed, it had access to everything on your phone: it could remotely record you, film you and it would gain access to any device your phone was paired with.

However, the most recent scandal surrounding Pegasus is that it has developed into a zero-click hack. This means you do not need to click on a dodgy link for the program to be installed on your phone. The hack is widely disseminated into devices via apps, most notoriously WhatsApp. Essentially this means that if you have WhatsApp on your phone, you could be spied on, which is really creepy. It has since emerged that at least 50,000 high-profile people – by which I mean the likes of politicians, diplomats, human rights activists, civil liberties activists and investigative journalists – were put at risk because of this spyware.

This is military-grade interception and spying software that was developed by a private corporation but sold to governments around the world, including many that have very questionable track records when it comes to respecting the human and civil rights of their citizens. In fact, the regimes that typically bought Pegasus were the ones that did not have particularly technologically evolved or effective intelligence agencies. The Russians, the

Chinese, the UK and the rest of Western Europe do not necessarily need this sort of software from a corporation, but there are plenty of paranoid regimes in Africa, Latin America and the Middle East that do, because they do not have the same access to state-level actors.

In this sense, corporations almost take on the mantle of spies. They are like digital mercenaries, offering malware and spyware for hire to those who can afford their services. Interestingly, the NSO Group claimed that they carried out background and ethical checks on all those they sold Pegasus to, and that they did not sell to regimes that would use it to target activists. However, when you look at who purchased the spyware, you can see the likes of Saudi Arabia and Morocco on the list.

The additional aspect we also now need to consider, in light of the Ukraine/Russia war, is the decentralisation of such tech tools and attacks. Groups of activists, hacktivists and individuals who in the past might have been considered criminal hackers by the powers that be have now been unleashed on both sides of the war. This technological genie cannot be pushed back into the bottle. Looking forward, these individual actors will present a threat to the usual power players such as national intelligence agencies, corporations, and even supra-national organisations such as the EU or UN.

A WEB OF SPY PHONES

In 2018, journalist Jamal Khashoggi, a prominent Saudi Arabian dissident, walked into the Saudi Arabian consulate in Istanbul, Turkey, never to be seen alive again. He had an

(*continued*)

(*Continued*)

appointment to collect the paperwork he needed to remarry. His fiancée, Hatice Cengiz, waited outside for him to reappear, but he never walked out of the consulate. Instead, he was brutally murdered and dismembered within the Saudi consulate, murdered by an elite hit squad sent by Saudi crown prince Mohammed bin Salman.

Four days after his brutal murder, Cengiz's phone was hacked using the Pegasus spyware. She was far from the only person in Khashoggi's circle to be targeted in this way. An investigation by *The Guardian* and other media[4] found that Khashoggi's wife, Hanan El-Atr, was targeted by the Pegasus spyware, as was Khashoggi's friend and former director general of the Al Jazeera television network, Wadah Khanfar.

The investigation found that a number of Khashoggi's associates were targeted for surveillance using the spyware in the wake of his death, including his son, Abdullah Khashoggi, and a Palestinian-British activist, Azzam Tamimi, who was also a friend of the journalist. Even the Istanbul chief prosecutor, İrfan Fidan, appeared on this list of possible surveillance targets earmarked by clients of the NSO Group.

Without forensic analysis of all the phones in question (and there were many more of Khashoggi's associates on that list of potential targets) it is impossible to know whether they were successfully hacked. However, the mere fact that they appeared on a list of potential targets speaks volumes about how Saudi Arabia (a known client of the NSO Group) was trying to gather intelligence and steer the investigation in the wake of the journalist's murder.

Pegasus spyware has been used to target not only individuals, but also governments and politicians. As recently as 2020, Pegasus made its way to Number 10 Downing Street and the Foreign Office. It appears that data from Number 10's computer network was exfiltrated after someone with a phone that had been hacked using Pegasus connected to the network, while the Foreign Office found evidence of that hacking. We can only guess at what sensitive information may have been accessed and by whom. The French government has also been targeted, with traces of Pegasus found on the phones of five current French cabinet ministers in 2021.[5]

This is a terrifying prospect. We do not know whether it is a foreign power, a rival intelligence agency or an individual actor who is behind these hacks. Regardless of who is responsible, we can be sure that whatever data or intelligence they gathered as a result will be used for nefarious purposes or to exert undue influence over democratic governments.

This is where the dark side of the corporate world collides with that of the dark side of government agencies. But corporations working alongside the spooks are not the only threat we face from this prong of the Dark Triad.

CHAPTER 5
Corporations

When it comes to the corporate world, there are two areas in particular that should concern us the most (although there are several others that I also explore in this chapter). These are, first and foremost, the monopolistic nature of many large corporations within the tech space and elsewhere. In relation to the big tech corporations, my biggest concerns lie in their willingness to do the work of nation-state actors (refer back to the previous chapter for more details of Snowden's disclosure of PRISM); and their aggressive activity to take over and buy out smaller, up-and-coming companies. You do not have to look hard to see that these huge corporations aggressively take over organisations that might present (1) a threat to their business model or (2) an attractive and viable alternative to what they are doing.

If you are reading this as a small business owner who has ambitions of growing in the tech space, or disrupting in any industry for that matter, this is certainly something to carefully consider.

To understand how we have reached the point we are at today, with the big corporations having such unwieldy power (particularly where intellectual property [IP] is concerned), it is important to understand when, why and how these companies started to rein in people's free access to information. I would like to start this chapter by going back to the late 1990s and the rise of the first dot-com bubble.

Nothing Comes for Free . . .

If you were alive (and old enough to use the internet!) in the 1990s, you may well remember Napster, which was a company that allowed people to download music for free. The big record labels and music moguls were unhappy about this and, particularly in the United States, regular people like you or me were being arrested just for downloading a few of their favourite songs on the grounds that they were infringing upon copyrights. If people were prosecuted, their sentences were often wildly disproportionate to their "crime". This was one of the first cases of big businesses in the entertainment industry seeking to keep a vice-like grip on the twentieth-century business model that, until that point, had helped them generate vast profits.

Let's skip forward just a few years to the early 2000s and the emergence of The Pirate Bay. This was a peer-to-peer file-sharing site that was developed in Sweden by Peter Sunde, Fredrik Neig and Gottfrid Svartholm. Movies and music were among the files that were shared on the site (although these were far from the only content on The Pirate Bay) and due to the vested corporate interests, the American government in particular aggressively pursued The Pirate Bay's founders.

However, the three men who established The Pirate Bay were very technically savvy and they mirrored their site all over the world. It was a bit like a virtual whack-a-mole: whenever the Americans tried to shut it down, it would just pop up somewhere else. Eventually, the governments and corporations won the war, with all three founders fined and sent to prison for supporting and facilitating copyright infringement.

These two cases alone demonstrate just how far we have strayed from the original, utopian idea of the internet, which was for people to have somewhere they could freely access information online. By the late 1990s, the internet had very much become controlled by corporations. Over the years, this has metastasized into controlling not only copyright in relation to entertainment, but IP on a much wider scale, encapsulating the likes of pharmaceutical drugs and even foodstuffs.

I have worked extensively with the Law Enforcement Action Partnership (LEAP) over the years, which is a non-profit organisation that advocates for drug policy and criminal justice reforms that will make communities safer. International bodies carried out research that found that over 80 per cent of the world's population were denied access to basic painkilling drugs because they were patented by Western corporations and therefore too expensive or subject to restrictive policies around their use (particularly where opioids are concerned).[1]

We have seen a similar pattern emerging during the Covid-19 pandemic, with the large pharmaceutical companies patenting vaccines and thereby limiting their supply to the rest of the world. Even though these companies should be rewarded for their work on developing medications and vaccines, the patents they are issued last for so long that the application of patents ahead of the global good, penalising developing countries, becomes indefensible – in the case of the Covid-19 pandemic, I would say that it is indefensible full stop for corporate profits to be put before global health. In fact, in withholding vaccines from the rest of the world, we are only shooting ourselves in the foot because this leaves more scope for variants of the virus to develop. Anyway, I digress.

When it comes to foodstuffs, there are corporations that have taken to patenting minor alterations they make to the genome of basic foodstuffs like rice, which then allows them to make farmers in regions such as Africa and Asia pay for a licence to grow rice every year, because the new plants will not self-seed.[2] This is a really insidious practice and it highlights everything that is wrong with IP, in that it is trying to preserve a twentieth-century business model that protects corporate interest.

Even when you look at IP in relation to musicians and authors, the legislation around copyright has never been about protecting the artists and their work. Instead it is all for the benefit of the intermediaries (aka record labels and publishers).

One of the stories I find most shocking in relation to IP protection, however, is that of Aaron Swartz. Before I talk specifically about Swartz, it is important to understand how the world of academic research and publishing works. In simple terms, academics (professors/researchers/lecturers) at universities in the West need to produce research papers to build their careers. These research papers need to be peer reviewed and, only then, are they allowed to be published.

However, there are only a few publishing houses within academia and they not only cost a lot of money to run but also charge a lot of money for people to access their information, which is scientific and academic research. Swartz was a brilliant tech student at MIT in the United States who believed that this model was wrong and that this kind of information should be publicly available. He therefore accessed the MIT database and downloaded a vast amount of academic research with the aim of making it freely available.

He was caught by the FBI and threatened with 35 years in prison for copyright infringement. At the age of 26, with this threat and the prospect of a trial ahead hanging over him, he committed suicide. Why have I told you his story? Because I want to help you see that IP does not only relate to music and films, but also to so much more that affects our lives. IP covers pharmaceuticals, academic research and even basic foodstuffs. All of this explains the first step of how this outdated twentieth-century IP business model was being applied to the internet and aggressively enforced.

The second step comes with the rather weird case of Kim Dotcom. . .

Blurred Lines

Kim Dotcom emerged from the German hacker scene and, having swum in those waters, I know that he is seen as a bit of a rogue. He set up a site called Megaupload, which is essentially a secure version of Dropbox where you can share all kinds of things, be they family photos and business documents or illegally downloaded movies and music. The point is, because Megaupload was so secure, you could not determine what content was being shared via the platform.

This is where it gets a bit weird and we stray from corporations enforcing copyright laws to state-level interference and a classic example of the global corporatist agenda. The action happened one quiet morning in 2012. By this time, Kim Dotcom was living in a mansion in New Zealand, but early one morning his peace was abruptly shattered when his property was raided by an FBI SWAT team. I am talking black helicopters, people rappelling down ropes – the works.

When the raid started, Kim Dotcom fled to his panic room (because all tech millionaires have a panic room!) thinking he was going to be assassinated. In actual fact, he was going to be arrested by the FBI at the behest of the American media corporations who said he was infringing copyright and IP laws. This was when it became apparent that a state-level actor like the FBI could be subverted and pressured by corporate interests in America to try and take out someone who they saw as an opponent. Megaupload was immediately taken down in the wake of the raid and Kim Dotcom, at the time of writing, is still fighting his extradition to the United States.

When you think about it, it is terrifying that a law enforcement agency like the FBI can be used in such a way by corporations for what turned out to be an illegal operation that invaded another country's sovereignty (because New Zealand did not agree to the operation and was not even warned about it).

Incidentally, Kim Dotcom has set up a new file-sharing site, Mega, which does not deal with any US hosters, domains or backbone providers, as well as changing how it operates to avoid another highly illegal US takedown.

All the examples I have shared here go to show how corporations have become involved in all aspects of the internet and how they are working for total control of the internet. This is why the corporations and intelligence agencies are the first two prongs of the Dark Triad. On the one hand you have the corporations trying to restrict and close down what information we can access and on the other you have governments trying to see what we are accessing.

The Ongoing Issue of Net Neutrality

Thrown into this mix we also have the ongoing debate about net neutrality, which for the time being has been maintained, but how long that will last is anyone's guess. For years, it has been like a bleeding sore that governments keep picking at; do they keep net neutrality, ensuring that everyone has equal access to the internet, or do they give preferential treatment (and faster internet connections) to corporations that can pay for it?

This is a debate that has been going on since the noughties, when corporations first started lobbying various governments and international organisations to try to ensure that businesses in certain industries – such as banking and the media – were able to access a high-speed connection. This was highlighted by Rop Gonggrijp in his talk at the Chaos Computer Club's 2005 event,[3] when he revealed that the giant corporations were closing down the open-source market and, at that time, starting to lobby against the concept of net neutrality. It is a debate that has continued ever since.

The Battle Between Lobbyists and Government

As well as having ties with intelligence agencies, the corporate world also does its best to exert influence over governments and encourages politicians to introduce laws that are entirely in their favour, but not anyone else's. The Transatlantic Trade and Investment Partnership (TTIP), which was at one point being negotiated between the EU and the United States, is a case in point.

Lobbyists pushed hard for the TTIP to be adopted, which would have been applied across the whole of the EU and would have opened the door for corporations to sue sovereign governments for the loss of potential future profits if those governments introduced laws that banned any of their products. For example, if the EU introduced a complete ban on genetically modified (GM) crops, any of the companies that produce GM seeds could sue those governments for damages.

When the TTIP was being discussed between 2013 and 2016, MPs were not even being allowed to examine the law they were being asked to rubber stamp. Only a few key people, who were heads of relevant committees, were allowed to read the legislation, such was the secrecy surrounding it; and these people had to visit an underground room where they were not allowed to take notes or make any recordings in order to pore over the 500+ pages of the TTIP.

Thankfully TTIP never made it through in the UK or the EU, but this is one of those revenant laws that, no matter how hard you try to kill it off, seems to be resurrected time and again under a slightly different guise in each iteration.

The Trickle-Down Effect

I appreciate that all of these stories sound very high-level and deal predominantly with the unhealthy sway that large corporations have over governments around the world. However, we all need to be aware that these big businesses are actively pressuring and lobbying governments to develop laws that are in their corporate interests.

This naturally brings us into areas like software and hardware, which suddenly brings the ramifications of these laws that focus on corporate interest into our own homes. Just think about how many people in the West now have "smart" homes. But these "smart" systems can easily be hacked, both by hackers and criminals (I share some true stories about exactly this in the next chapter). We have doorbells that can spy on us, "smart" toys for our children that could be hacked into by paedophiles and, of course, we have our "smart" speakers and home assistants that are always listening, providing vast amounts of data about our lives to these big corporations. The questions to ask yourself are whether you trust these corporations to protect and look after your data, and whether you trust them to do the right thing in terms of human rights.

Corporate Threats and Opportunities for SMEs

If you are running a small to medium-sized enterprise (SME), you might not immediately see how the behaviour of these big corporations has a knock-on effect on your organisation. However, the impact of what they do, particularly through their lobbying of politicians, could well trickle down and impact how you run your business.

Just imagine what might happen if a key piece of legislation you rely on for your business operations gets changed as a result of corporate pressure. Or whether new regulations being introduced within your industry could have a negative impact on your company. Even if this is not a concern of yours right now, that is not to say that a year or two down the line things will not be changing.

This can and does happen, a good example being Klout.com, which was a data aggregator that allowed you to see your impact across social media online. It was taken offline when the GDPR legislation was introduced in Europe. The business was completely blindsided by the new regulations and what they meant for their operations.

That said, the behaviour of the big corporates also presents opportunities for SMEs, particularly those operating in the tech space. More and more people are waking up to the potential threats to their privacy as a result of the technology that we use in our daily lives, and this presents genuine opportunities for businesses that are able to produce tech products that are of high quality and stamped with a degree of authenticity that allows consumers to trust them.

One example is Dutch smartphone manufacturer Fairphone, which not only allows you to configure your smartphone the way that you want to, but also allows you to repair or replace any components at any time. This makes these devices much more sustainable, because you do not need to upgrade your whole phone in order to improve one element of it or add new functionality. Fairphone uses open-source hardware and as much open-source software as it can on its devices (I talk more about the many benefits of open-source in the final part of this book).

The founder of Fairphone told me that, while not all their software is open-source at present, they are working with French company EOS to ensure that they have a totally open Android operating system on their phones (rather than a Google

Android operating system). This means that users no longer have to rely on old apps that may no longer be supported, that might not be patched, or that could be more vulnerable to hacking.

When it comes to hardware, however, most of our technology is made in China and one of the little-known Snowden disclosures from 2013 is that, since 2008, all hardware, from laptops to USB cables, has had backdoors built into it, which could be exploited by state-level actors, corporations or criminals.[4]

This presents a real challenge for any small business that wants to build that platform of authenticity and security around their products. If you are sourcing your hardware components from China, know that they will all have these backdoors built in. However, if you source your hardware components from elsewhere, for example within Europe, the cost is typically considerably higher and this presents a barrier to entry for consumers, who may not want to or be able to afford to pay more for their tech devices.

Phil Zimmerman, who created PGP encryption, also set up a business producing secure smartphones, which was known as Blackphone. This business manufactured all its hardware components in Switzerland, which made the products very expensive and limited their uptake. Blackphone has since ceased trading. The message is that, if you want ultimate security, you are going to have to pay for it. But again, this circles back to the argument that you should not have to be rich to be able to keep your privacy intact.

What if you are not in the business of creating new tech products? Does any of this matter to you? As I have explained earlier in the book, the fact that these backdoors into our technology exist means that there are vulnerabilities that can be exploited, whether by criminals, spies or the corporations themselves.

It can be easy to think that if you are a small business that you will not be a target, but that is not necessarily the case.

Certainly if you are offering a product or service that could compete (albeit a long way in the future) with one of the big corporate giants, the chances are you will be targeted or bought out long before you become a serious threat. If you are relying on using software and other services from the big tech companies, and have licences with them for your business, you are leaving yourself exposed to an attack. There are ways to lessen this particular threat, such as opting for open-source operating systems and software, which I discuss in much greater detail in the final part of the book.

There is one further area that can pose a particular risk to businesses, which relates to top-tier domain names (.com, .org, .info and .net). The US government claims hegemony over these domain names because they are based in the United States. This means that, without warning or reparation, the US government can take down any website within one of those domains. Your website, and all its content, can literally disappear overnight and be deranked and disappeared by Google. All because, under US law, the government has the right to do this globally.

What does this mean to a small business? Simply be aware that this is a genuine threat. By all means get a .com domain, because that is how everyone will find you, but make sure your site and all its content is backed up somewhere safer.

The trend I have discussed for taking mastery of IP, whether in the tech space, pharmaceuticals, food, academia or elsewhere, will ultimately stifle innovation and limit what small businesses are able to create for the good of all of us.

Sustainability: A Key Battleground

Sustainability is becoming an increasingly important conversation for businesses in all industries, but the tech industry is coming under the spotlight more and more frequently. Many people are concerned about what is happening to our planet and how it is being stripped of its resources, with an appalling toll not only on the environment, but also on the local communities that live in regions that are rich in the assets that the tech industry needs.

Most people do not think about where their tech comes from, who makes it, what rare minerals and heavy metals are used for its manufacture, or how the mining of those minerals and metals affects the health of those working to extract them. Many individuals in the West are just keen to upgrade their smartphone each year, without a second thought for the resources required to create it. However, this is beginning to change and a growing number of people are waking up to the significant sustainability and environmental issues surrounding the manufacturing of the technology products we all rely on.

This is another key area where SMEs can differentiate themselves, by focusing on the ethical manufacture and sourcing of technology products. As a business, you can take an ethical stance about how you operate, both in terms of the physical products you may sell and how you operate online.

Different Approaches to Our Tech Evolution

One of the areas I find particularly fascinating is looking at how different countries have approached the technological evolution that we all faced at the same time. If you explore how China and the West have all evolved, for example, you can see they have taken very different paths despite having access to, broadly, a lot of the same technology.

In the West, we live in a capitalist society where our access to the internet and technology is decentralised from government control and held in the hands of corporations. This means that in the West we have a soft underbelly that can be exposed and make us even more vulnerable to the predators that are out there, be they corporations, nation states or criminals. This is how capitalism makes us, as individuals, more vulnerable. It is also important to note that there are different shades of the West – Europe and the United States are very different beasts in terms of how their societies operate and what support is provided from the state. However, it is the Washington consensus that has allowed this form of corporatism to spread.

By contrast, China has taken a centralised approach, where the state controls access to both the internet and technology. However odious you believe China's approach is, it raises an interesting question – is it in many ways more secure because you only have one vector of attack (from the state), rather than this Dark Triad that also introduces the corporations and criminal elements?

We also have to ask what lessons we can learn from the different approaches taken in different nations. Is Western hegemony or Chinese hegemony any better or worse than the other?

What Does the Future Hold?

If we look far into the future (although actually not that far!) there are going to be some incredibly important questions we need to ask about how corporations can protect us in an online world. This might sound like something out of a sci-fi story, but there is a very real possibility that within the next 70 years or so, humans will be in a position to upload their consciousness into the digital realm, effectively enabling us to live forever.

If this becomes reality, the questions would be, Who would own the hardware into which we are uploaded? Who would be able to spy on our consciousness? Who would have access to our consciousness? Who would have copyright or IP ownership over us? Who could tamper with our consciousness? All of this comes down to one very important point: Who do we trust to manage this and protect us from all three prongs of the Dark Triad, namely the spies, the corporations and the criminals?

As fanciful as this may sound, there is going to be an interesting test of some of these points in the next few years with the emergence of the metaverse. Businesses, both large and small, are going to be encouraged to embrace the metaverse and to use it to connect with customers, particularly those who are digital natives and who are most likely to embrace the concept. And the concept is cool; it almost feels like an early step towards transhumanism where you will be able to shape your identity in a way that has not been possible before. That is a good thing.

However, there are still questions over who will potentially own your individuality, who can watch it and who could potentially tamper with it in an online environment like this. These are key questions we have to ask *now*, if we are to protect our online identities for the future. It is only by doing so that we can ensure that our rights are protected in this new online environment.

CHAPTER 6
Criminals

Before I dive into some of the main criminal threats we face, I would just like you to take a moment to look around the room you are in as you read this book and make a mental note of the technology you can see. Do you have a smart TV on your wall? Is there a smart speaker in the corner? Maybe a laptop on the coffee table? How about a baby monitor? I am sure your phone is within easy reach, too. Simply make yourself aware of all the technology that is around you; I come back to this later in the chapter.

The Sophistication of Digital Crime

As I mentioned in the previous chapter, conventional criminals have become a lot more sophisticated in how they use technology to target both individuals and businesses. You might still get the odd email from a Nigerian prince asking you to click on a link, but most of us are savvy enough not to fall for that these days! However, more and more criminal hacks are carried out using emails that appear to come from someone we know (look at the CEO example I shared in Chapter 4), or they do not require you to click on a dodgy link at all.

When the NSA cyber weapons cache was stolen by the Shadow Brokers and leaked online, it gave criminals access to a host of backdoors that allow them to carry out zero-click hacks, where they can simply access our phones or computers using the apps we have installed. These kinds of leaks not only allow the criminals to use the vulnerabilities that are exposed, but also allow these hacks and viruses to mutate.

These vulnerabilities can also be reverse-engineered by criminals to get into people's phones, access all of their personal data and, in some cases, to threaten them. We all live our lives on our phones. Most of us in the West bank using an app on our phones; it is incredibly convenient to be able to pay with our phones, travel, and carry out all kinds of activities with this single device. The problem is, these devices are highly insecure pieces of hardware. Of course, if criminals can gain access to such vulnerabilities, they will exploit them.

What Is the Scale of the Problem?

There is a growing gulf between the literate and illiterate when it comes to technology. Those who can code do code, and their capabilities are accelerating far beyond those who simply consume technology. I am talking about not only the wealthy tech billionaires, but everyday people who are waking up to the threats we all face and taking steps to protect themselves. What I would like to do is begin to bridge that gulf by helping more people become aware of the threats they face simply by using technology without considering the security implications. This gulf is set to become even wider with the advent of quantum computing and the introduction of more and more artificial intelligence (AI), which is a topic I return to later in the book.

For now, it is important to point out that the level of cyber crime is only increasing. Ransomware attacks are a case in point, where a criminal hacker gains access to online systems, usually of a business, and shuts the business out of its data systems. They then hold the data and system hostage and only allow the business access again on the payment of a ransom, which is usually a huge sum, often payable only in cryptocurrency.

Even though ransomware attacks have been happening for over a decade and there is greater awareness of what they are and how they are executed, there are still a significant number of these kinds of criminal attacks happening each year – in 2021, 37 per cent of all businesses and organisations were hit by ransomware, with recovery from a ransomware attack costing businesses an average of $1.85 million. Ransomware attacks cost the world $20 billion in 2021, with this figure projected to increase to a staggering $265 billion by 2031.[1] As I mentioned, the criminals are also getting more sophisticated in how they inject the malware into these large corporate systems.

WannaCry is one of the most notorious recent examples of a ransomware attack, but there have been many others. The problem for businesses is that, with so much work being conducted via computer networks and on the cloud, it only takes one person to make a mistake for an entire network to become infected and compromised. If you are using your personal computer for work purposes, and your device is connected to your employer's network, you could also find you are personally affected by this kind of cyber attack.

Why Should This Matter to You?

Ransomware attacks targeted at businesses might have an impact if you work for the company involved, but as an average individual you could be forgiven for thinking criminal hackers would have no interest in you, or would not be able to target you in a similar way.

As I explained in Part One, even if you have not done anything wrong, that does not mean that you want every aspect of your life to be in the public domain. Your financial and health records are a case in point, and it is this kind of information that criminals are likely to target you with. Another tactic that cyber criminals are increasingly employing is gathering data about your online searches and viewing habits, usually related to pornography.

Watching porn (certainly in Europe) is not illegal nor should it be anything to be ashamed of. Having video sex with your partner is also nothing to be ashamed of, and many people have certainly tried that since the arrival of the Covid-19 pandemic! However, this can leave you wide open to blackmail and ransomware attacks by criminals, who will threaten to post compromising photos or videos of you, or your browsing habits, online. Do you want that information to be out there for the world to see?

Some criminals will even send speculative emails without having your information. However, if you receive an email asking for money or else they will release X, Y, Z information about you publicly, you have to be very confident in your online activities to know that is a speculative attempt.

I once received just such an email, from someone calling herself Susana Peritz, telling me that she had hacked my *email, planted malware on my computer and then filmed me getting my jollies while watching "interesting" porn online. This "Susana" demanded that I pay $1,000 via Bitcoin or (shock!) have my alleged pleasures shared with my acquaintances. This email caught my attention because the subject line included a very old password, which was attached to an email account I had not used for a decade or more. Had this all been true, any malware would have been planted on a very old (and now defunct) computer. As a result of my time working for MI5 and then on the run from the spies with David, I always keep my computer camera lens covered. I can also safely say, hand on heart, that I have never watched online porn. "Susana" had tried to shake down the wrong paranoid ex-MI5 whistleblower! However, if you were to receive such an email, would you have my certainty in your position and therefore be confident enough to ignore such a request?*

Stalkerware is another criminal threat that we all need to be aware of in an increasingly digital world. I mentioned in Chapter 1 that corporations have received backlash for listening in to people's lives via their smart speakers, but if a corporation has that capability then so too do criminal hackers. Having someone hack into your smart speaker and eavesdrop on you is bad enough, but there is an even darker side to this kind of criminal activity.

Researchers have shown just how easy it could be for paedophiles to hack into smart toys or smart watches that children use, allowing them to take photos and videos of them or in some cases even learn their location and send them text messages pretending to be a parent or other person they trust.[2] The implications of this are truly terrifying.

It is not only the toys themselves that pose a danger, but also the toymakers who fail to secure the data they collect from children using these smart toys. VTech was fined $650,000 in 2018 for failing to adequately protect children's privacy and for breaking US laws about how data from children was gathered. The company's Kid Connect app, which was bundled in with many of its toys and used by almost 650,000 children, was hacked by a security researcher, who thankfully took their findings to the press rather than using the data for nefarious processes. The hacker discovered they could access not only personal data, but also an internal database holding encryption keys that, had they been used, would have allowed an attacker to access photos and audio files uploaded by parents and children.[3]

Criminals can hack into smart doorbells to see who is coming and going at your home, and therefore when your property is empty. Your smart TV, laptop, webcam or computer can all be hacked to listen in or video you while you go about your business in your own home.

Perhaps one of the most terrifying examples comes from the reported hacking of Ring home security devices in 2019. There were multiple reports of hackers accessing these home security devices to watch people, including unsuspecting children, in their own homes and even to talk to them. There were also reports of hackers tricking people into letting strangers into their homes by showing them a video of someone they knew standing at their door.[4]

Many of our cars nowadays have smart keys, which can be cloned by criminals allowing them to steal high-end cars. There are so many ways in which the smart technology we are inviting into our lives because we think it is cool and convenient can be used against us.

LOVEINT: Where Criminals and Law Enforcement Meet

LOVEINT is a practice whereby law enforcement, government or intelligence officials abuse their access to information that is afforded to them by their position. People who hold these roles have access to a great deal more data about us than the average person. There have been a number of cases where someone in that position has abused their power, position of trust and the system by using it to spy on people of interest to them in their own lives. This includes a current or ex-partner, someone they have had a casual relationship with and even new friends they are getting to know.[5] Essentially they use this backdoor access to cyber stalk this person, whether by accessing confidential records about them or even hacking into devices in their homes (like their smart TV) so they can watch them.

> *Take a moment to think back to all the smart devices you noticed when I encouraged you at the beginning of this chapter to look around the space you are in. I am willing to bet that, all of a sudden, your smart speaker, smart TV and even your mobile phone do not seem quite so innocuous now.*

Awareness Is Essential for Protection

One of the biggest challenges when it comes to combating digital crime of any nature is that the vast majority of people are simply unaware of the potential vulnerabilities in the software and hardware they use every day. Until we all become more awake to these possible threats, it will be very difficult to effectively mitigate against them.

Many of the threats we face in a criminal sense have been exacerbated by the Covid-19 pandemic and the significant rise in the number of people working from home. Pre-Covid, people would go into an office for work, where they would assume (rightly or wrongly) that the business would use antivirus software and that there were IT policies and procedures in place to keep them safe. Now, more people are working from home and often using their personal devices to do so.

Another issue with relying on your company's IT team to keep you safe from criminal threats in the digital world is that many corporate IT teams themselves are struggling to keep up to date with what these threats are. Digital and online technology is evolving so rapidly that it is a genuine challenge to keep pace with that change and this means that often corporate policies and the technology they use to protect the business and its employees from online threats are out of date.

One of the most important factors, therefore, in mitigating against the criminal threat is education. The more people are aware of the vulnerabilities that exist, the more they will take appropriate steps to protect not only themselves, but also the business, online. This will create a self-perpetuating loop of greater knowledge, where individuals have the knowledge to protect themselves more effectively in a digital world and, by association, protect businesses more effectively as well.

Essentially, more engaged staff and citizenry as a whole will provide greater protection for all of us from the criminal threats that exist. There is a wonderful example from nature of how this can work incredibly effectively. Certain bee species have learnt that if hornets attack their hive, the best defence is to swarm around a hornet and vibrate, which raises the hornet's body temperature and kills it. Working together makes it possible to neutralise the threat much more efficiently.

When David and I went on the run from MI5, we purchased a laptop in a random shop, paying cash, to make it as secure as possible. However, we were still paranoid as it had a built-in camera and microphone. We knew that the intelligence services could use keylogger strokes to pick up passwords and things like that. We were therefore incredibly careful about what we used that laptop for, we covered the camera and so on. But nowadays it is not only intelligence agencies you have to worry about being able to access your information. Criminals can hack into all of our devices more easily than you and I would like, and as the amount of technology in our homes has exponentially increased, so too has the risk we are all exposed to.

BIG BROTHER IS WATCHING . . .

Around the time of the Snowden disclosures in 2013, a German MP decided to see just how much he was being tracked on a daily basis, both via CCTV and his smartphone. He wanted to see how integrated the technology was, so he spent a day walking around the city and then accessed the data about his day. He received images of exactly where he was throughout the day – and this was almost a decade ago; things have become much worse in the years since.

(*continued*)

(*Continued*)

London is known for being one of the most surveilled cities in the world, with one of the highest per-capita number of CCTV cameras. With the introduction of more and more facial recognition technology, these systems are wide open to abuse or unintentional misuse. Facial recognition is still in its early stages and it is far from accurate. Let's say you look like someone who is in the criminal database. You are waiting for a Tube in London and you get picked up by the CCTV and misidentified as this criminal. This could result in your arrest, even if only until they realise they have the wrong person.

Just like the German politician, we are all being tracked every day through the GPS on our smartphone. As I said at the beginning of this book, it is up to you to carry out a risk assessment and decide what you are prepared to trade off in favour of convenience.

How Is Cyber Crime Being Tackled?

Law enforcement agencies are trying to tackle cyber crime, but in all honesty they have limited success. Takedowns of these cyber crime groups often have happened as a result of human intelligence, rather than cyber intelligence. The EncroChat case is a good example. EncroChat ran as a legitimate business, promising its users an "end-to-end security solution to guarantee anonymity". Its smartphones cost around £1,500, were Android-based and not only had encrypted chat apps installed by default, but also had their cameras, microphones, USB data ports and GPS sensors removed.

In 2020, Dutch and French authorities infiltrated EncroChat after four years of work. Data from the 60,000 users worldwide was harvested and shared via Europol. Of those users, around 10,000 were in the UK and using the system to communicate about a range of criminal activities, from drug and human trafficking to illegal weapons sales and ordering hits on rivals. When EncroChat realised it had been infiltrated, the company sent a message to all of its users urging them to discard their devices before shutting down and disappearing without a trace. However, law enforcement was already moving. The resulting series of raids by UK police resulted in 746 arrests, as well as the seizure of more than two tonnes of drugs, 77 firearms and £54 million in criminal money.[6]

However, as technology is evolving so rapidly it is incredibly difficult for those working in law enforcement to keep up, which is why it is so important that we all take responsibility for our online security and do everything we can to protect our own security and privacy. You cannot rely on the government or corporations to keep you safe, so use the power of your wallet to apply pressure to corporations and use your voice to apply political pressure to democratically elected officials to take cyber crime more seriously and do what is necessary to protect our privacy and data. Until then, do everything in your power to make your digital devices more secure.

Cyber Criminal or Hacktivist? Who Decides?

Another inescapable element of the cyber crime debate is the actual definition of a cyber criminal. Some of the examples I have shared here, such as paedophiles hacking children's toys or baby monitors and capturing and then disseminating images, are obviously criminal. There is no question or doubt of that. However, there are other areas that are less clear-cut.

I am sure you are aware of the saying, "One person's terrorist is another person's freedom fighter" and it is no different in the cyber criminal/hacktivist space. Anonymous is a classic example. They have long been branded criminals because they have hacked into governments and corporations for what they deem to be political causes. As I write this book, Anonymous have been praised throughout the West for hacking the Russian state to protest against the war in Ukraine. All of a sudden they are heroes, when just a matter of months ago they were branded criminals.

To add another cliché into this mix, "The victor is the one who writes history". In the future, this could mean that Anonymous get a free pass because of their activities in relation to the Ukraine war. But that does not change the fact that many members of that group have been arrested and imprisoned over the years because of their activities hacking corporations and governments in an attempt to expose corporate and government crimes.

From my personal experience of blowing the whistle and going on the run, I can tell you that there is something of a Robin Hood complex among those of us who are considered criminals by some and freedom fighters by others. For the most part, those who go down the route of hacking to expose wrongdoing believe they are doing so for the greater good. In exposing information in this way, they want to benefit society. It is a "robbing the rich to feed the poor" mentality.

There are numerous examples throughout history where in order to change the law you first have to break it. Look at the Suffragette movement, or those campaigning against racial segregation in the United States in the 1960s. Galileo Galilei, Nelson Mandela, Malcolm X and Rosa Parks are among the many figures who were persecuted during their lifetimes only to be vindicated later. How these people are perceived now, as trailblazers and activists who changed the course of society, is very different from how they were perceived by many in their own time, when they were considered criminals. History looks on them favourably; the question is, will it do so on the hack-tivists of today?

CHAPTER 7
Media Control

Media control is all about defining the narrative, because whoever controls the narrative defines the process or progress of international relations, whether in relation to elections, referenda or wars. We have seen this time and again throughout the conflicts in the Middle East – the invasions of Afghanistan and Iraq, the Libyan civil war, the war in Syria, and the ongoing situation in Israel and Palestine – but as I write this it is escalating to an entirely new level with the war being waged by Russia in Ukraine.

We are seeing how the impact of cyber war has resulted in the use of the likes of deep fakes to wage propaganda wars not only in the old media, but also on new media platforms such as social media sites.

DEEP FAKE: VOLODYMYR ZELENSKY

At the beginning of March, just a week after Russia launched its invasion of Ukraine, the Ukrainian government's Center for Strategic Communication warned people that Russia was likely to use deep fake videos in an attempt to encourage

(*continued*)

(*Continued*)

Ukrainians to surrender. Just two weeks later, the first such attempt appeared.

A fake video of Zelensky appeared on social media sites, including Facebook, YouTube and Russian social network VKontakte, and Telegram. A still of the video and written description was also posted on Ukrainian TV channel TV24's website. In this video, "Zelensky" called for Ukrainian troops to lay down their weapons and surrender. This particular deep fake was not the most expertly executed, with the Ukrainian president sounding different from normal and his head appearing to be too big for his body.

TV24 was quick to react, announcing it had been hacked and taking the deep fake and its written summary down within minutes, with other social media sites removing the video just as quickly.

Mere minutes after the TV24 announcement about the hack, the real Zelensky was on the offensive, posting his own Facebook video in which he denied he had asked Ukrainians to surrender and calling the deep fake a "childish provocation".[1] This deep fake was easy to spot and debunk due to its poor quality and the high profile Zelensky has had on the world stage since Russia invaded his country. Yet not all deep fakes will be so easily and quickly identified and removed.

The concept of deep fakes is not new. The technology that allows this kind of content to be produced has been evolving for about 20 years. Colin Powell, former US Secretary of State, saw a demonstration of deep fake audio technology in around 2003–2004 from the country's intelligence agencies. They presented him with an audio recording of his "voice" saying something contradictory to his normal position. He was apparently taken aback and wanted to know how anyone would be able to tell this audio was not him.

While this technology may have initially been restricted to intelligence agencies, we know that they are unable to keep it under wraps. It has, as we have seen, become much more accessible and therefore mainstream. It is trickle-down technology, starting with the spooks and finding its way into criminal hands and to those who have nefarious purposes.

The Evolution of Media Control

However, to understand the level of media control we face today, and that we will face in the future, it is important to understand how media control has evolved over the decades, first in the mainstream or legacy media and then into an online environment that also encompasses social media.

Full disclosure: I grew up in a family of journalists. My grandfather and father were editors of the local papers in Guernsey, where there is a surprising amount going on due to its status as a tax haven. From a very young age, they taught me that you need to speak truth to power; that is what the media is for. You cannot have an informed citizenry if you do not have a free media that is legitimately speaking truth to power. I grew up with this mindset and I certainly think that, to a certain extent, it informed my decision to blow the whistle when I was working at MI5. However, it was after blowing the whistle and seeing how my former partner David was treated, or should I say maltreated, by the media that I truly realised the power the mainstream media has to spin the narrative.

> *David's legal case lasted for seven years and, of course, the media were very interested and there was a huge amount of coverage. I was shocked at how he was represented, though, because I would see him giving interviews and then see how what he said in those interviews would be twisted and changed in the mainstream media. This process horrified and fascinated me.*

> *He was on trial in the Old Bailey for a breach of the Official Secrets Act and, of course, he was convicted because there is no legal defence under the Official Secrets Act (1989). However, I made a mitigation plea to the judge after he was convicted but before his sentencing, where I explained what our motivation was for whistleblowing and what had happened. In his final ruling, the judge said that David had not put any agents' lives at risk by blowing the whistle and that he had not done it for money.*

> *However, the next day all the newspapers led with headlines along the lines of "Shayler sells agents' lives down the river for money". It was the polar opposite of what the judge had said, and I had seen most of those journalists in the courtroom when he delivered his final ruling.*

This was when I really started thinking about how the accurate reporting of a story could be manipulated to such a degree behind the scenes and as a result I started having conversations with a lot of the better journalists I knew to learn more.

Manipulating the Press

Through these conversations, I learnt about the various ways in which the old media can be, and is, manipulated. The first is through what I describe as the "carrot" approach. This is where you might have two people who were at college or university together, for example, one of whom now works for MI5 or MI6 and the other who has become a journalist on Fleet Street. The two of them meet up for lunch, it is all very friendly, and the intelligence agent might initiate a conversation with "I've got a wonderful story for you . . ." or they might go in with, "I've seen you're covering this story, but it's not quite accurate. Could you spin it like this . . .".

Ultimately what happens is that the journalist is offered more stories and potentially more scoops to help them build their career. This secret charmed circle behind the scenes interlinks all the way up the command chain. There are connections between editors, senior spooks, politicians and the owners of the media too.

Another approach taken by the intelligence services to manipulate the media is to plant fake stories as well as to massage detrimental stories in favour of the spies. Back in the 1990s, there was a division called Information Operations (or I-Ops) within MI6 that served precisely this purpose. Although the name will have changed after David Shayler outed it, I have no doubt that it still exists.

FAKE STORIES AND ULTERIOR MOTIVES

In 1995, the *Sunday Telegraph* fell hook, line and sinker for a classic MI6 propaganda operation, where an MI6 operative, at the behest of I-Ops, spun an elaborate fake story about Saif al Islam, son of Colonel Gaddafi, and his involvement in currency fraud. Hapless spook hack Con Coughlin ran with it and the fake story's publication and subsequent paper trail was used as a convenient pretext for the authorities to deny Saif al Islam entry to the UK at the time.

In 2002, Saif won his libel case against the *Sunday Telegraph*, receiving a grovelling apology from the newspaper while refraining from asking for "exemplary damages", which he would almost certainly have won. It has since emerged that just months after the court case, he was welcomed to the UK to study at London School of Economics and afforded MI6-backed protection while he was here.[2]

Finally, you have the hard powers or the "stick" – the laws that can be used against journalists and news outlets to prevent them from publishing stories. This includes injunctions, public interest immunity certificates (which are government injunctions), super injunctions (where the press cannot even say they have been gagged) and libel laws. There are also terrorism laws, which can be used to try and provoke journalists into revealing their sources. Former journalist Chris Mullin, who is known for exposing the wrongful conviction of the Birmingham Six in the 1980s, was at one point being pressured into revealing his sources from that investigation, using exactly these laws.[3] He refused to reveal his sources, risking prison in doing so, and finally won his High Court battle in the UK in March 2022.[4]

In addition to these legal powers, there is also the Official Secrets Act. Contrary to popular belief, this act does not only apply to spies, espionage and whistleblowers; it also covers journalists. At the time of this writing, a journalist who breaks a story relating to national security would face two years in prison if it can be proved that they damaged national security by publishing a story, just as a whistleblower faces two years in prison per charge, automatically, if they break the Official Secrets Act.

As I write this, the UK government is trying to coalesce all the different elements of the Official Secrets Acts from 1911 onwards into something called the Espionage Act, which will make journalists liable for up to 14 years in prison just for doing their jobs and reporting a scandal that involves national security. Of course, this tariff will also apply per charge to the whistleblowers emerging from central government, the military, diplomacy and intelligence – the very people who are most likely to witness the most heinous crimes.

These are all intimidation tactics designed to discourage journalists from reporting on stories that the spooks would rather stay hidden. We have seen examples of how the intelligence agencies in the UK use intimidation to try and prevent journalists from publishing stories in the past. One of the most high-profile examples was in 2013, when *The Guardian* was forced by GCHQ to destroy the hard drives it had containing the disclosures from Edward Snowden. Even though the paper's editors told the authorities that they were not the sole custodians of nor the only recipients of Snowden's files, the intelligence agency still used these tactics to try and frighten the journalists off.

In the years that I have been giving talks on this subject, many investigative journalists from all over the world have approached me and told me that the situation is even worse in their country. The UK is often held up as a beacon of democracy, but behind the scenes the strings, levers and pulleys of control are very powerful when it comes to the traditional mainstream media.

The Power of Press Reports

In the wake of Donald Trump's election win in the United States in 2016 there was a great deal of reporting about alleged Russian interference in the election, although I put the "Russia-gate" years from 2015 until he left office in 2021. The way in which the media was being spun around his tenure was truly fascinating.

Even before Trump took office, the CIA and FBI were working hard to sow the seeds of the "Russia-gate" narrative. They positioned Russia as "the enemy" that was working to undermine US democratic processes, while providing very little evidence to back up their assertions. As Robert Parry revealed in an article in 2017, the US intelligence community wanted to discredit both Trump and Clinton, because it did not approve of either candidate.[5]

Following Trump's election, the CIA certainly didn't hold back any punches. A slew of fake news stories followed, from the evidence-lite assertion that it was the Russians who hacked the Democratic National Committee (DNC) emails and leaked them to WikiLeaks (something Assange denied in a highly unusual move) to stories about Russia hacking the voting computers. This quickly morphed into the "Russia hacked the election" narrative that was consistently wielded by the media (and stoked by US intelligence agencies) throughout Trump's presidency.[6]

A lot of fingers were pointed at Facebook, because of their microtargeting activities, specifically around people's views. This evolved into the Cambridge Analytica scandal a year or two later. There was a shaping of the whole narrative that suddenly everything pointed to Russia and that any hacking must come from there.

THE CAMBRIDGE ANALYTICA SCANDAL

Cambridge Analytica was set up as a political consulting firm, with the financial backing of Robert Mercer (a secretive hedge-fund billionaire and Republican donor). Steve Bannon, who is of course closely tied to Trump, was the company's vice president in 2016. Most notably, Cambridge Analytica worked for the Trump presidential campaign, during which time it harvested the data of up to 87 million Facebook users.[7]

The concept behind Cambridge Analytica's work was to combine "information operations" (as used by intelligence and military organisations) with big data and social media, allowing the company to produce sophisticated psychological and political profiles of millions of people. As whistleblower Chris Wylie explained in an interview,[8] Steve Bannon believes that politics is downstream from culture, so to change politics you first have to change culture, which is what Cambridge Analytica set out to do.

In order to achieve this, the firm built a Facebook quiz that allowed it to collect and exploit the data of not only those

(continued)

(*Continued*)

who took the quiz, but also the friends of the people who took the quiz. All of this happened without any of the people whose data was being harvested having any idea it was happening.[9] This is despite Facebook's terms not allowing data gathered in this way to be sold or used for advertising.

There is much debate over the level to which Cambridge Analytica's activities influenced the outcome of the 2016 US presidential election. The firm has also been investigated for its role in working with the "Leave" campaign in the UK's referendum on Brexit.

Wylie revealed, somewhat terrifyingly, that Cambridge Analytica's parent company SCL Group has worked with both the British government and the US Department of Defense to carry out "counter-extremism operations" in the Middle East. By 2017, when Bannon was Trump's chief strategist in the White House, SCL was pitching for work with the Pentagon. As Wylie put it in his interview with *The Guardian,* "The company has created psychological profiles of 230 million Americans. And now they want to work with the Pentagon? It's like Nixon on steroids".

However, this form of manipulation has been happening for decades. One historical example comes from the UK in 1923: the publication of the Zinoviev letter in the *Daily Mail* ahead of a general election. The Labour party was in power and touted to win the election, but this letter allegedly written by Grigory Zinoviev, the head of Communist International (Comintern) in Moscow, implied that the Labour government was receiving money from Moscow. Labour lost the election by a substantial margin, putting the Conservatives in power. Years later, it emerged that the Zinoviev letter was a forgery.

There are similar stories from the United States, too. For example, in the 1960s and 1970s the CIA ran Operation Mockingbird, where they had plants in editorial rooms. A similar approach was taken in the UK by the intelligence agencies, and I have been told by several journalists in the UK that not only were they aware that this was going on, but they all knew who the spooks were within their ranks and either avoided them or used them as they could.

Modern Media Censorship

In the modern era and the age of the internet, it might appear that it is easier than ever to share stories; but it is also easier than ever to make them disappear. I have seen this first-hand with articles I wrote and interviews I gave during the turbulent years following mine and David's decision to blow the whistle and go on the run. I have paper copies of all of these articles and interviews, but when I go to the digital version of those publications, they have all been removed. It is very surreal to see all the other content from those dates still available, yet anything relating to me or David has disappeared. It is like we have been airbrushed out of history.

One of the reasons I have been a huge supporter of the Wiki-Leaks model is that it acts as a direct conduit between the information and the people who should be able to read that information. The idea is to remove the blockages between the people of the world and the information that they need to know. One of the challenges as a whistleblower is knowing whether you can trust the mainstream media to publish what you share. The answer is that you cannot trust this to happen. This is why WikiLeaks is a trailblazing, revolutionary organisation, because it let the genie out of the bottle technologically, and it has been persecuted severely for that precise reason ever since.

There are now countless examples of modern media and internet companies creating blockages in terms of what we, as individuals, are able to access online. As I write this, Russia is waging a brutal war in Ukraine, which I utterly condemn. Extraordinary steps have been taken to curtail Russian media during this period, with TV and radio stations being taken offline; Facebook, Instagram and Twitter are all stating that they will take down any posts that relate to Russia's propaganda about the war.

I am not an apologist for the Russian state; I think it is odious and brutal, and is becoming more so every day. However, when you look objectively at the stance of the social media companies, you have to ask how they are going to develop the algorithms to remove such posts, or whether they have already developed these algorithms that are, in the most basic sense, going to censor what we are able to see and read.

If you take this capability out of the context of the war Russia is waging against Ukraine, it is a terrifying development. The fact that it can and has been done is worrying for the future evolution of free access to information. It is a demonstration of the narrowing of people's options of what they are allowed to read and learn from and therefore prevents people from making their own balanced evaluations of situations by using different news sources.

Another example of how media outlets are being censored and discredited in this way comes from the PropOrNot website, which is a shadowy, anonymous organisation that published a list of websites that it claimed were "Russian propaganda". This list included WikiLeaks, *Consortium News* (a highly respected online digital news platform in the United States), and the World Socialist Web Site (WSWS), a huge site with a big readership.

In the week following the publication of the PropOrNot list, WSWS lost approximately 70 per cent of their traffic,[10] because they were deemed to be "useful idiots associated with Putin". At the same time, there was a pincer movement with Google, which started deranking these websites. This meant that, even if you searched for "World Socialist Web Site", it would not appear on the first page of the search results, but would be buried several pages further on in the search results. The fact is that most of us click on the first search item Google provides, or certainly within the first few. Most people are not going to trawl through 15 pages of search results to find information. Independent sites like this should not be stimatised or "disappeared" in this way, and it takes them a lot of effort to recover, if they ever can.

This deranking element, which is linked to political and media agendas, is very concerning, because it means any information that does not conform with the orthodoxy of the social media company that is holding it or that has algorithms managing that information can just disappear through the cracks. It is another way of airbrushing history. This is not only relevant to politics, but also to corporate and individual reputations.

Striking the Right Balance

There are cases where social media companies *should* remove content; for example, any posted material that is criminal should be taken down. Paedophilia, snuff videos and terrorism absolutely fall into this category, although sadly through the Dark Web this kind of content will always find a home. In this instance, where something is already deemed to be criminal and widely stigmatised, of course the social media companies should have rules that allow them to take that content down.

However, the issue comes when what is being posted is not criminal but is an opinion or a perspective. Is a particular freedom of expression acceptable politically, or might it anger certain sectors of society? The question then becomes who is going to police that and who is going to be the arbiter when it comes to social media?

We already know that social media creates echo chambers, because what we are shown in our feeds is so highly targeted to us as individuals. This means that we only see similar opinions and articles that support our views, rather than anything that might challenge them. Many people do not realise just how much the information they are being presented with is being "censored" in this way, because they think of themselves as consuming information from the vast domain that is the internet.

Particularly in the West, discussions have become much more polarised and there is a liberal orthodoxy that is feeding that. In the United States, almost 50 per cent of Americans voted for Trump. In the UK, just over 50 per cent of people voted for Brexit. Both tribes have been vilified for doing so by this liberal orthodoxy. Personally, I do not agree with either Trump or Brexit, but that is not the point. The point is that it is becoming easier for liberal tech bosses in Silicon Valley to drown out those voices and only promote what they see as the more liberal side of things. However, I would see that approach as being utterly illiberal.

Ultimately this comes down to the personalities of the people who are controlling these big tech giants, because they are now controlling the global narrative. The problem is who is controlling *them*. We have seen time and again that these global corporations cannot be held to account by governments, whether in the UK, the United States, or elsewhere. The bosses of these companies have appeared in front of Congress and the UK Parliament, where they have been asked to give evidence to an inquiry or even just to pay their fair share of taxes, and every time lawmakers struggle to hold those organisations to account.

COUNTERMEASURE: FUTURE-PROOFING INFORMATION WITH THE FREEDOMS INDEX

I am a director of the World Ethical Data Foundation, which focuses on R&D in this space. We are working on something called the Freedoms Index. Its purpose is to future-proof information and it is initially being built to apply to human rights but once it is up and running can be applied to any form of information. The idea is that any information held in the Freedoms Index will be protected from being deranked or disappeared, or from being censored.

We are using Blockchain technology, where you can stamp the packets as they go through and authenticate the information,

(continued)

(Continued)

because this is a very good way of ensuring that this information cannot just "disappear". We have also developed a numeric way in which to stamp this data, which breaks down all linguistic barriers. For example, if you are interested in a human rights issue in Chile, but you do not speak Spanish, you can still search for and find the information you are looking for. This will break the Google search methodology, which is very linguistic-based, as well as removing the threat of information being deranked.

Who Really Has Control?

What we all need to bear in mind is that the inherent bias in any of the algorithms that are used come from whoever creates the algorithm or, moving forward, the AI. With AI in particular, once it starts being used and it takes off, it learns and replicates by itself, and that means it can go off on many different tangents – even those who have created it lose control of it. There are many studies that demonstrate inherent bias when it comes to programming. One in particular is the real-time facial recognition software that has been used by many police forces in the UK and that has been an egregious failure, particularly in South Wales and in the Metropolitan Police, over the past few years. The reason is that most of the people doing the programming are youngish white-ish men and the software has been found to be appallingly inaccurate when it comes to the facial recognition of anyone who is not white. The Metropolitan Police has admitted that there is a significant gender bias within its facial recognition technology, with women misidentified at higher rates than men. Various studies from the United States also show that

Black people are more likely to be misidentified than those with light skin.[11]

This potentially means that you could be going about your business, perfectly innocently, only to be picked up on a CCTV camera in real time, misidentified and as a result approached or even arrested.

In one worrying example shared by Big Brother Watch, a 14-year-old Black schoolchild, who was wearing a school uniform, was surrounded by four plainclothes police officers who questioned him after he was incorrectly identified by the facial recognition system. The police officers asked for the boy's phone and took his fingerprints before realising they had the wrong person. The incident was caught on film and shared by Big Brother Watch,[12] showing that the child who was misidentified was scared and felt he was being harassed by the police.

As I write this, the launch of the metaverse is fast approaching and this online realm is predicted to become where we live most of our lives, rather than in the real world, within the next decade. But do we want to, first, put ourselves in a position where our lives are owned by any of the corporations behind the metaverse, and, second, how will this affect the narrative we are being fed? If we are in an even more immersive environment, what will that mean in terms of how we absorb information and the nature of the information we are shown?

We also need to consider how these new platforms can be abused by those with nefarious purposes in mind. In Chapter 6 I highlighted some examples where smart technology, including children's toys, have been hacked by criminals. There are also some incredibly disturbing cases emerging from some of the current metaverse platforms. In 2022, a 13-year-old girl in the United

States was kidnapped and repeatedly sexually assaulted by a 33-year-old man after meeting and chatting to him on the gaming platform Roblox.[13] These platforms need to do better at safeguarding us and our children in these online spaces.

Proper controls need to be put in place within the metaverse by the organisations operating in that space. As individuals we need to make these controls a condition of our interaction with such platforms. It is clear that, based on previous attempts by governments to hold big tech firms to account, expecting international bodies and governments to hold these organisations accountable and police them effectively is not realistic.

When the internet was launched, the utopian concept was that it would be somewhere we could access all human information, a space that would provide a free flow of information and a place where people could both learn and express themselves freely. However, this concept has been lost as a result of first the corporate takeover of the internet and the way the platform is now controlled in large part by monopolistic corporations, and second the intervention by state-level actors from various nations. This has pushed us towards more of a dystopian rather than a utopian future.

That is not to say there is not still hope. As I explain in the final part of this book, there are steps we can take at every level of society to help steer our course back towards the utopian ideals the internet embodied in its early days and away from a future that sounds like it has come straight out of the pages of a dystopian sci-fi novel.

CHAPTER 8
Cyber Warfare

What Is Cyber Warfare?

I would define cyber warfare as the use of a cyber weapon for political means, rather than just criminal. Cyber warfare is not just about trying to make money, it is about trying to make a political point or win a political position.

Therefore, a cyber attack is not, by definition, an act of war. However, a cyber weapon or a tool being used by either countries or people thinking they are acting in the best interests of their country can potentially become warfare. When this happens, it has the potential to get hot very quickly. We have seen this evidenced in the attacks going back and forth between Russia and Ukraine since 2014, when Russia annexed Crimea.

There are also two strands to cyber warfare, one being the information warfare that is happening between countries and the other being weaponised cyber tools that have been used to attack critical infrastructure, such as energy or transport networks, in other countries. These are targeted cyber attacks on the infrastructure that is necessary to keep a country functioning. This is not dissimilar to the approach the Provisional IRA took in the UK in the 1990s. They wanted to attack military and infrastructure targets, but they did not want to attack and harm civilians, both because it was not the right thing to do and because it was counterproductive from a reputational perspective. Cyber warfare takes a similar approach in that it goes after a country's infrastructure while causing minimal personal or civilian harm.

Most countries, and certainly those in the West, have become overly reliant on technology for running their basic civil infrastructure, making it an attractive target. In addition, most of that tech is built on old, outdated systems, hardware, programs and proprietary software, leaving it wide open to all kinds of cyber threats; again, a prime example being the WannaCry ransomware hack.

One of the biggest challenges is that most people and indeed politicians do not even know that they need to consider these issues when it comes to protecting their country's infrastructure. They do not question how the infrastructure functions and assume that it will continue to run as it always has. In doing so, they are leaving their countries wide open to these sorts of threats. The threat of a cyber attack is not new, but it has accelerated significantly in recent years.

What Infrastructure Is Vulnerable to a Cyber Attack?

In short, anything. Even having a closed system that is not connected to the internet does not guarantee protection (as we learned from the Stuxnet attack carried out on the Iranian nuclear facility), although it does make it more challenging. The most extreme risk in terms of cyber warfare would be a hack involving military systems. It is not beyond the realm of possibility to see a scenario in which a nuke might be set off, although I sincerely hope that we are entering the *reductio ad absurdum* argument here and that this is not possible.

What we have seen very recently in Russia's war with Ukraine, however, is an internet communication satellite being targeted and taken out. In February 2022, Viasat reported a "cyber event" that had caused a partial network outage, leaving thousands of satellite internet users without online access in Ukraine, as well as in Germany and other European nations. This was more than likely a Russian cyber attack, with disruption to services reported to have lasted more than two weeks.[1]

It is also conceivable that a cyber attack could take out plane navigation systems. We have seen examples where railway networks have been shut down. Shipping company Maersk was hit by a malware attack in 2017,[2] affecting its operations and ability to deliver goods. If you think about all the ways in which we rely on the internet in the modern world, you can see there are many known unknowns in relation to cyber warfare's sheer scale of capabilities to cripple entire societies. We are reliant on technology for supply chains for our food, our electricity supply, even our water supplies. Your electricity supply being cut off does not just mean no lights and electronic devices. It also means you will not be able to flush your toilet, or pump fresh water into your home. All these little things we take for granted can very easily be hit through cyber warfare.

If a cyber attack were to knock out an energy supply network, or to knock out the satellite navigation of the tankers that bring fuel into Europe, the whole infrastructure of society would collapse within as little as a week. Threat models have been developed for this very situation. Even a decade ago, I remember reading about how we have become so inherently reliant on data and the networks in Europe that, if they were taken down, civilisation as we know it would fall apart within a week to 10 days, certainly in big cities.

We have known about the threat of cyber attacks on infrastructure for at least 25 years. In 2022, for the first time certainly in Western Europe, we have seen how such cyber attacks can be used, and it is frightening. Many politicians have certainly realised how imperative it is that they consider such threats in the wake of Russia's war on Ukraine.

Russia has attacked Ukrainian infrastructure both through physical bombardment and cyber attacks since the beginning of the conflict, with internet connections in Ukraine severely disrupted as a result. Elon Musk's SpaceX Starlink internet service has provided a lifeline and ensured that many regions of Ukraine have remained connected to the internet. Over 10,000 emergency Starlink receivers were donated initially and have proved crucial to Ukraine's communication network since the start of the invasion.[3] There have also been reports of Russian hackers targeting Ukraine's water and electricity supplies. This is now part of modern warfare.

However, it is not only state-level actors who are taking such actions. We are seeing an aggregation of individuals coming together to support one side or another in a conflict and attacking infrastructure or corporations. The situation is out of control. It is no longer in the hands of a state or a corporation; it is in the hands of a swarm of people on the internet, whose allegiance may change, and that is a huge shift. We have to ask how corporations, governments and even we as individuals can deal with this. We all rely so heavily on this infrastructure that these are not questions we can ignore.

What I have talked about so far are the physical threats to our infrastructure, and, by association, society, that could result from a weaponised cyber attack as part of a cyber warfare campaign. The other aspect of cyber warfare lies in the control of information.

Cyber Propaganda

Cyber propaganda is just as important as the caches of cyber weapons that nations are developing, because it is the propaganda that enables the manipulation of entire populations and I do not think you can distinguish between how a population is manipulated into believing something and the use of weaponry to enforce that belief.

If you look back at World War II, the Nazis not only built up a huge stash of weapons and rearmed, even though they were not legally allowed to following World War I, but they also created a huge propaganda campaign, which was absolutely disgusting. Several years ago I visited Cologne in Germany. Most people will tell you that Cologne Cathedral was the only building left standing in the city after the carpet bombing in the war, but that is not quite true. The other building that survived was the Gestapo HQ, which has now been turned into a museum.

The Gestapo held and tortured prisoners from across Europe in the cells at the HQ. There is a lot of graffiti on the walls left by those imprisoned during this period, which is very moving to see. But what I found astounding was the exhibit on the upper floors of the museum, which is dedicated to the propaganda that the Nazis created and disseminated about the Jews in the runup to World War II. This propaganda was used to manipulate the country's entire population, which is why I do not believe you can separate the propaganda from the weapons used in war.

This tactic was used not only in World War II. We have seen the same thing happen ahead of both Iraq wars, particularly the 2003 conflict when what has now been shown to be a fake dossier about Iraq[4] holding weapons of mass destruction was used to justify the invasion that led to millions of people suffering in the Middle East.[5]

What we are seeing now on the internet with data is a faster, more assured and more effective way of spreading this propaganda. In the previous chapter I mentioned the microtargeting carried out by Cambridge Analytica on Facebook, although microtargeting is happening increasingly on Twitter as well. In addition, most of these big systems are based in the West, specifically America. What has happened since the beginning of Russia's war on Ukraine is that Russia has been cut off from all of these systems to prevent the Russians from spreading their narrative that they are carrying out "a special military operation to liberate the people of Ukraine". In terms of the information war, I would therefore consider this to be a cyber propaganda weapon that is just as necessary and just as effective as launching cyber weapons themselves.

Cyber propaganda is often about preparing the ground before unleashing a real-world attack, as well as being about creating or promoting a narrative that encourages people to commit violence or an act of war. Again, this is what happened ahead of the second Iraq war and also the NATO invasion of Libya in 2011. Cyber propaganda therefore becomes a cyber weapon, because it is used to weaponise people and encourage them to act on a narrative that has been created for political gain.

I talked about media control in the previous chapter, but this also feeds into the cyber propaganda narrative. If you look at Russia, for example, the media has always been very tightly controlled but it is reaching new levels of control and censorship in 2022. Since Russia launched its war with Ukraine, it has introduced laws where the penalties for any dissent are severe and this is significantly restricting news reporting in the country to ensure that everything follows the narrative being shared by the Russian government.

Who Are the Key Players in Cyber Warfare?

We know that the UK and United States are both active in the cyber warfare field and, indeed, as you have already learnt, agencies like the NSA, FBI and CIA have seen their cache of cyber weapons leaked onto the dark web (or, in the case of the CIA, details have been shared more responsibly via WikiLeaks).

In 2020, General Sanders, a cyber war expert in the UK, said there were 60 countries around the world that had major cyber warfare capabilities.[6] He went on to name just four (China, North Korea, Russia and Iran), but we can make an educated guess about many of the others. Ukraine is another key player in this global web of cyber warfare, and it would be fair to assume that most Western European countries, along with Australia and New Zealand, also have considerable cyber warfare capabilities.

The National Cyber Power Index 2020, produced by the Belfer Center,[7] put the United States at the top of a list of the world's countries that have high levels of capabilities and intent within the cyber field. China was second, the UK third, Russia fourth and the Netherlands fifth in this index. While the United States, UK and China (along with France and Germany) were named as nations with higher capability and higher intent, Russia (along with Iran, Israel and the Netherlands) fell into the higher intent, lower capability bracket, meaning they have either not publicly disclosed their capabilities or do not currently have the capabilities to achieve their cyber goals.

China, I would argue, have historically been more interested in industrial espionage and using their capabilities to build their own economy.

Cyber Warfare Goes Beyond Nation States

When I use the term cyber warfare, I do not only mean national cyber warfare. As I have already explained, the cyber weapons that have been developed can just as easily (and frequently are) turned on companies as well as on nation states – the WannaCry ransomware attack being one of the most prominent. Companies therefore need to think about how they can cyber-proof themselves because, as you have learned, the caches of cyber weapons that have been developed by various national intelligence agencies have not been kept secure. On top of that, the cyber weapons that private intelligence corporations have developed have now also found their way into criminal hands.

The problem is only escalating as a result of the war in Ukraine. Now, groups of Ukrainian hackers that had previously worked with Russian hackers to attack targets in the West are turning on one another for political reasons and spreading each other's code and software online. This simply means that even more of it will fall into criminal hands and that it can therefore be used to target individuals as well as corporations.

These "mafia gangs", for want of a better term, of hackers have been encouraged or at least tacitly supported by nation states when they are acting in their interests, one example being during the Russia-gate scandal in the United States and the alleged subversion of the US democratic process by these criminal gangs in Russia and Ukraine. At that time (2015–2016), these particular gangs were acting together to work against the West. What we are now seeing, however, is these gangs turning on each other as the war between Russia and Ukraine escalates.

Previously, there was a tacit understanding between Russia (the nation state in this example) and the hacker gangs in Russia and Ukraine; they were united against a common enemy (the West). However, some of these criminal gangs have gone from being on one side of the conflict to the other, in that they previously supported Russia and are now supporting Ukraine (and, by association, the West) following the Russian invasion, and they have political motivations. These criminal gangs have been (and will continue to be) used by nation state actors for their advantage and plausible deniability.

This presents a number of problems from a cyber security perspective, most notably that these criminal gangs that were once working together are now leaking details of each other's cyber weapons and tools online. This means that even more malware, worms and cyber weapons are open to being mutated and utilised by other criminals elsewhere in the world. Businesses in particular need to be particularly aware of these ever-evolving threats and, if they have not already, need to think seriously about building serious cyber defences. This is a topic we have been discussing at the World Ethical Data Foundation and we are taking further steps to protect our own data and websites as a result of the emerging threats we are seeing at the time of writing.

The speed of escalation is unprecedented, which means corporations can become very easy targets for criminal hacking gangs if they do not keep on top of their cyber security. As we know from earlier chapters, many are not taking this threat seriously. With quantum computing and quantum cryptography on the horizon, it is more important than ever that businesses and individuals take the cyber threat seriously. If you are using out-of-date systems or technology, you are already leaving yourself wide open to a cyber attack. There is a driving need for anyone heading up a big organisation to ensure that the technology they are using is, at the very least, patched and ideally upgraded to ensure that they are not using out-of-date technology that is more vulnerable to hacks.

Another solution to consider is switching to open-source software, which, while not invulnerable, is one of the least vulnerable options in terms of protecting your data and network because it is constantly being monitored, checked and upgraded to protect against backdoor attacks (which is how ransomware works).

Are We Prepared?

Cyber warfare is very much at the forefront of twenty-first-century conflicts. We might still have the tanks and the nukes, but those are weapons that very much belong to the twentieth century. Yet what still amazes me is that so many people (outside of cyber security circles) do not seem to have seen this coming.

As an individual you might wonder why you should care, but the reason is simply that we cannot rely on our governments to keep our data safe. We have already seen significant cyber weapons caches from Western governments falling into criminal hands and this will happen again. Companies cannot rely on governments to protect them from cyber attacks. It is up to us. We need to take our own stand and, as an organisation or an individual, we need to take that responsibility for keeping our data safe into our own hands.

As I said earlier, educating ourselves about the potential threats is one part of the puzzle, and this is exactly what you are doing as you are reading this. We need to be aware that these threats will continue to evolve and mutate as new technology and capabilities develop. We also need to be mindful about the predations of the state and consider how we will protect ourselves should, for example, more draconian laws be introduced in response to an ongoing conflict, or even another more severe global pandemic.

What is going on politically in the world has a bigger impact on us as individuals than you may initially think. For example, what happens if Russia retaliates against the likes of MasterCard and Visa for being excluded from the SWIFT payment system in Europe? Will your financial records be safe? Or if there is another global pandemic and, as with Covid-19, we are expected to carry our biometric and health data around on our smartphones, how can we protect that personal data in a digital environment?

The point is, if these tools and weapons are out there and the spooks cannot even protect them to ensure they do not fall into the hands of criminals, this makes all of us very vulnerable and that is why, as individuals, we have to take responsibility for protecting our data and lives online.

PART III
Solutions

Although I may have painted a somewhat bleak picture of the current state of affairs in places, we are not irrevocably set on a track toward a dystopian future, where surveillance is rife and privacy feels like a far-removed concept rather than being the reality of how we live our lives.

We have a choice about whether we educate ourselves and contribute to a future where technology empowers us, or whether we fail to take action and allow technology to control us. There are steps all of us can take as individuals to improve our privacy and protect our human rights in a digital world. Businesses, too, can take measures to help build this free and equitable future.

Finally, governments need to recognise the vast possibilities that lie before us where technology is concerned and introduce policies that educate and empower, rather than seeking to dictate and control how we access and use technology in the future. In the coming chapters, I share details of the various steps, or countermeasures, we can take at various levels of society, starting with us as individuals, then exploring the corporate world, and finally looking at what governments and elected officials can do to help set us on the right path for the future.

I also invite you to think about the future we are building for the generations coming through behind us. What kind of world do we want them to live in? Technological developments that are expected in the next decade will mean that possibilities that once only existed in sci-fi literature could become reality. Only by educating ourselves about technology and ensuring that the next generation is properly educated about how to code and program that technology to protect their own privacy and rights can we start to bring about the cultural shift necessary to set us on a trajectory toward a future that more closely resembles utopia than dystopia.

CHAPTER 9
Individuals

Many of the threats I discussed in Part Two might sound somewhat distant, but as you learnt in Chapter 6, when I explored some of the criminal activities going on, I found that these threats are much closer to home than many of us would like.

So, as an individual, or someone who wants to protect their family, what do you need to think about in terms of what steps you can take to protect yourself from some of the threats I mentioned in Part Two? You could be forgiven for thinking that there is not much you can do given the global nature of the threats, but that is far from the case. Each one of us can take responsibility for our online privacy and security, and there are tools out there to help all of us become safer online.

Before you decide which steps you want to take, however, you will need to carry out your own risk assessment based on your situation. If you live somewhere with a very repressive regime, you may choose to take more of the steps I am about to share with you than someone in a comfortable, notional Western democracy, for example.

The level of risk you face will also depend on your activities. If you are just an average citizen going about their business, you may feel comfortable with a lower level of protection than, for instance, a journalist who needs to protect their sources, a lawyer who wants to protect legally privileged conversations or a doctor who wants to protect their patients' health records. There will be another level of threat assessment altogether if, for example, you are an activist in a country where you need to protect not only yourself but also the people you are talking to.

Each person's threat assessment will be different and that will likely dictate which of the tools and suggestions I share here that you choose to implement in your own life. The suite of tools you choose to use will depend on how paranoid you think you need to be. There are tools available to protect your communications, your computer, your knowledge and your data. The key is, first, in knowing that these tools exist and how you can access them.

Of course, the more tools and levels of security you add, the more of a pain this can seem, which is why it is important to carry out your own threat assessment, decide what and who you want to protect, and then find the best way to do that.

Awareness Is the First Step

Simply being aware of the various threats that are out there is the first, and probably the most important, step. Without knowing that you need to protect yourself and your family online, you will do very little to guard against the risks. Now that you know about the morass of threats we all face, you can start to take action in the most appropriate way for your lifestyle.

You do not know what you do not know, and many people simply have no idea that there are legitimate and very effective alternatives to the closed proprietary software that comes installed on your devices, whether it is a PC, a laptop or a smartphone. This brings me to the first of the steps I would recommend for protecting your online privacy: switching to open-source software.

The open-source community provides a very powerful tool for individuals to protect their privacy online by developing a raft of different technology and making it freely available at this level. These developers do, of course, have to make money, and they do this not by charging individuals but by offering their programs and software to corporations and governments for a fee, which may also give them access to additional functionality.

Countermeasure #1: Open-Source Software

I will take you back to 2007, when my understanding of tech was fairly primitive. I could use a computer, but I had no idea how it worked, who was controlling it or who had access to what – and this is a former spy talking! This was when I found out about open-source software, and I remember having a moment when the scales fell from my eyes. I had never even thought there could be an alternative to Apple or Microsoft software. And therein lies the challenge: you do not know what you do not know, so if you are unaware there is an alternative to these major tech players, you will not even ask the question. I have used open-source software ever since I discovered it in 2007.

Learning about it was a revelation. You do have a choice and you do have control; you do not simply have to stick with what comes preinstalled on your device.

As I have explained, closed proprietary software is vulnerable to having backdoors installed in it, which leaves users open to zero-day hacks and stalkerware. Nor do you always get timely security patches to protect your data, either because the software is no longer supported or because other players, such as spies, are holding back this information. These can be exploited by criminals, as well as by intelligence agencies and this is the greatest argument for switching to open-source software.

I find open-source software itself fascinating, because it has been developed and supported by a community of publicly minded technologists and geeks. The source code (as the name suggests!) is open, which means that anyone can review it. Every time a new open-source program is released, it is reviewed by this global community and, if there are vulnerabilities, they are not only picked up but also patched. These updates happen automatically, so you do not have to wait for these patches to be implemented. This is in contrast to the closed proprietary companies, which may not even know about some of the vulnerabilities in their code (as we have seen) and may therefore take longer to issue a patch.

There are various flavours of open-source software available. Personally, I use Linux Ubuntu, but there are others like Debian and GNOME that are also available. All of them are available to download and install, free of charge. They are free to use, freely scrutinised and freely protected by a global network of people who are interested in the principles behind the open-source movement.

Even if you do not know how to download and replace the existing operating system on your device, you will likely have a techie friend who can help! It is easier to do than you might imagine. I also find that open-source software is more user friendly than many of the closed proprietary software systems that are available. In terms of usability, it is certainly comparable to the closed proprietary operating systems, and the open-source systems are evolving all the time.

I come back to how open-source software can work for corporations in the next chapter, but it is important to note that open-source software is not necessarily free when it is used by businesses or governments, because this is how the software engineers involved make their living. However, from an individual perspective, it is an amazing project that works for the common good and gives people access to more secure and user-friendly software, free of charge.

When you are buying a new electronic device, you have two options when it comes to open-source software. The first, as I have mentioned, is to buy a device that comes with preinstalled closed proprietary software, which you then strip out and replace with an open-source operating system. The other option is to buy a clean device with no proprietary software installed at all. In fact, there is a growing market for this kind of device, where you are able to install whatever operating system and software you would like, without any preinstalled firmware and software to strip out.

Countermeasure #2: Open-Source Web Browsers

Another, potentially even simpler, step you can take to protect your online privacy is to use an open-source web browser. Mozilla Firefox is the web browser I recommend. As with other open-source software, it is free to download and to use.

Through Mozilla Firefox, you can download all sorts of open-source tools that can protect your web browsing to a certain degree, the most notable of which are ad blockers. You can search for the various options through Mozilla and choose the one that sounds good to you. This will simply mean that you will not see adverts when you are browsing websites, which I think makes for a much nicer online experience!

You can also download cookie control programs, called cookie crunchers, again through Mozilla Firefox. These simply mean that, once you leave a website, the cookies it has installed on your machine are "crunched" by the program automatically. This keeps your web browsing a little more pristine and it makes it much easier because your browser is not infested with all of these adverts.

You can install ad blockers and cookie crunchers on other browsers, such as Google Chrome or Microsoft Edge, but the reason I personally prefer Mozilla Firefox is that it was developed by the open-source community and has tried its best to adhere to those standards as closely and openly as possible. This is not about one option being better than another; it is about you having a choice and, I would argue, that using an open-source web browser offers you a fighting chance of having more privacy online.

There is also Mozilla Thunderbird, which is a free, open-source email application that you can download and use to manage your emails. One of the reasons I like it is that it allows you to download all of your emails and work offline, with your replies being sent in bulk once you are online again. I also recommend Mozilla Thunderbird because it makes it very easy for you to have access to email encryption programs.

Countermeasure #3: Email Encryption Programs

As I explained in Chapter 2, PGP email encryption is still considered to be the most secure email encryption program available (despite a human idiot a few years ago accidentally leaking some passwords). That incident aside, PGP encryption still has not been broken.

One of the reasons PGP encryption is so secure is that it is asymmetrical. What this means in practice is that you have a public key, which is out there and that anyone can access, and you have your own private key that will meet with it when you accept a linkage. As it is asymmetrical and you are the only person who has the private key, PGP cannot be cracked. Even Edward Snowden has stated that, pre-quantum computing, it is the most secure encryption program you can use. After the password leaks that I mentioned, all of us using PGP encryption had to cut new keys, but despite this it is still a secure system.

PGP encryption used to be incredibly difficult to download and install. I remember when David and I were hiding out in the farmhouse in France with only a dial-up internet connection that he managed to download and install PGP on his computer, which was an incredibly technical feat for a non-technical person. Nowadays, however, the likes of Mozilla Thunderbird have made it much easier – you simply click on a button, install a key and off you go!

Whether you feel this level of encryption is necessary for you will, of course, depend on your threat assessment. Personally I started using PGP encryption over a decade ago because, due to the people I was in contact with and emailing, it felt rude not to use a PGP encryption key. When I email a new contact, I even reflexively check whether they are using PGP encryption. While I do not expect everyone to use it, I prefer to have the option of using it and I believe using PGP encryption demonstrates consideration for others' privacy as well as for your own.

If you do not want the inconvenience of having to think about PGP, there are other options, including a number of very security-focused organisations stemming from the open-source community. ProtonMail is a good example, because it too has asymmetrical encryption like PGP. All you have to do is create a ProtonMail account, download the email system and you can get started. The basic version is free to download and simple to use. If you want additional functionality, you can pay to upgrade; there are also upgrade options for corporations and large organisations.

ProtonMail is an email system in its own right, much like Gmail or Outlook. If you want to have a ProtonMail account in addition to keeping existing email accounts, you can use an email management system like Mozilla Thunderbird to have all your accounts and emails in one place.

Countermeasure #4: Privacy-Focused Search Engines

Rather than using the default search engine on your computer, you can use a search engine that places a greater focus on privacy while browsing the internet. Examples of these kinds of search engines include DuckDuckGo and Startpage.com (which is what I use). I like to describe these as a prophylactic measure to protect your privacy online. Essentially these programs prevent the main search engines from seeing everything you are searching for without you needing to go into the Dark Web (which can be a scary place!).

While the likes of DuckDuckGo and Startpage.com do not give you full protection, they do allow you to search the internet without ads and cookies tracking you.

Countermeasure #5: Tor (The Onion Router)

Tor is another step up from the options I just mentioned, because it provides at least three steps of separation between what you are searching for and what comes into your computer. This completely anonymises your searches on the internet.

Tor is an open-source program and, much like the other options I have shared so far, has become much easier to download and use than it used to be. Installing Tor on your computer is as simple as visiting the Tor Project website and clicking "Download Tor Browser". The beauty of Tor is that you can switch it on and off, so you do not need to use it all the time if you prefer not to. It might simply be that you want to look up something that could be considered embarrassing, that you want to research a private medical issue, or that you want to talk to a fellow activist or journalist securely.

Countermeasure #6: Secure, Non-Proprietary Operating Systems

If you have a computer that you use for work that requires you to have a specific closed proprietary operating system and software installed, Tails is the answer for your personal privacy and security. Tails is a separate and portable operating system that you can have on a flash drive. All you do to use it is plug the flash drive in and log into your private operating system. Tails will bypass the normal login on your device.

This allows you to have an entirely different computer setup, one which is open-source and much more private and protected than anything offered by the closed proprietary programs. You can also use Tails in conjunction with Tor, allowing you to browse the internet much more privately and securely from your Tails operating system. You can install Tails from the Tails Project website.

Another option in a similar vein is the Qubes operating system, which is intensely privacy focused and was developed within the hacker scene before being made more widely available. You can download and install Qubes from its website. Like Tails, it is open-source, and the Qubes operating system allows you to divvy up the different areas of your computer. Essentially it allows you to compartmentalise your digital life, much as you would in your real life where you might share certain things with your family and friends, but not with your colleagues. It is a way to run different aspects of your life privately, without the need for multiple devices to do so.

Countermeasure #7: Virtual Privacy Network (VPN)

A VPN essentially disguises where you are searching from on the internet and it loses a record of what you search for. When you have a VPN installed, you can choose which country you want to search from. In fact, VPNs have grown in popularity in recent years as a way of accessing content on streaming services that is not available in certain regions.

Accessing different streaming services aside, VPNs are a first step in attempting to have more anonymity, and thereby privacy, when you are searching online. They are at the lower end of the spectrum in terms of security, but if you are just a normal person who does not have a high threat level, they are a good option to give you a greater degree of privacy and security, and one that is easy to use.

Digital Empowerment

Everything I have shared here is designed to empower you and show you that you have a choice beyond what you are given off the shelf when you buy new technology. Some of these solutions may be a bit much for you and your assessment of the threats you face, and that is fine. All I want to do in sharing these solutions is to show you that there are options out there beyond the closed proprietary software and to help you see that you have a choice.

If you start using the likes of ad blockers, you will even find that your online browsing experience is improved, because you will not be bombarded by adverts on every website that you visit. Far from making using the internet less convenient, I find that it makes it much more convenient and just more pleasant. For example, using Mozilla Firefox, with the free ad blockers I have downloaded, I can watch content on YouTube without any adverts. Just imagine being able to watch a video without having to wait for an advert to finish first, and without those horrendous "commercial breaks" popping up between videos. All of this is possible with open-source software and tools.

What About Smartphones?

Everything I have talked about so far has been very much focused on improving your privacy and security when you are using a laptop or PC. Of course, many of us use our phones for a lot of internet browsing these days, and plenty more online activity besides. It is, however, much more challenging to install open-source software on a smartphone than on a computer. That is not to say that smartphones that use an open-source operating system do not exist, but they are incredibly expensive, which puts them beyond the reach of most people.

I mentioned Blackphone in Chapter 5, which no longer exists as a smartphone manufacturer. The company has evolved into Silent Circle, which allows you to add encryption to an existing device, but you will still be using the proprietary operating software that comes installed on your device.

I use a proprietary smartphone and have simply accepted that my smartphone is wide open and therefore I do not trust it. So, if you want to have a private conversation these days, you have two choices: (1) accept that it can be listened in on, or (2) put your phone in the fridge (like the hacktivists in Berlin used to), as this creates a makeshift Faraday cage.

Even if you cannot change the hardware and operating system on your smartphone, you can still be mindful of the apps that you download and use. WhatsApp, for instance, is widely known to have been hacked. Signal, on the other hand (at the time of writing at least), is still relatively secure. This messaging service (which offers much of the same functionality of WhatsApp) was developed in the hardcore open-source developer community and was end-to-end encrypted.

However, in 2022 the company announced that, in order to scale up, it was going to have to close a couple of kernels (meaning that the entirety of the software would no longer be open source), which has concerned some of us who use it. This is, however, a good example of how the technology scene is constantly shifting and evolving, and also a timely reminder of why it is important to keep up to date with the latest apps and software being developed and released.

There are certainly companies out there that are trying to catch up to some of the big closed proprietary software businesses and offer open-source alternatives, particularly in the sphere of messaging and video meetings following the explosion of this technology during the Covid-19 pandemic. Some of the open-source software in this area includes Jitsi, Element and Matrix. These are more secure than their closed proprietary counterparts, but they are functionally not quite as good (yet!). There is always something new hovering over the horizon, so even if there is not a solution that suits you right now, that does not mean there will not be one in the future.

What About Smart Home Devices and the Hardware Itself?

My advice would be to not have them in your home in the first place! If, however, you already have a smart speaker or some other smart device (or you are set on getting one), the best advice I can give you is to change the code immediately and make sure you are not using the factory settings, because this is how most hackers can access these devices.

Hardware is a trickier beast. As I mentioned in Part One, one of the lesser-known Snowden disclosures is that, since 2008, all hardware that has been manufactured has had backdoors built into it. There are a number of factors at play here, from the fight for the mastery of the internet to China's economic aggression, but ultimately it means that no matter what hardware you buy and what software you use, there will be backdoors built into it.

As a result, there is still a huge market for pre-2008 laptops, particularly IBM ThinkPads, which were considered some of the safest devices because you could take them apart and reconfigure them to your own specifications, installing whatever you wanted. The issue, however, is that the functionality of this technology is much lower and you need to be technically minded to know how to set them up. This makes this approach highly inconvenient and therefore only something you are likely to consider if you believe the threat level to you and your family is high.

Privacy and Tech Can Go Hand in Hand

Ultimately, ensuring your privacy does not have to mean that you stop using technology or, indeed, that it makes using technology any less convenient. In fact, as you can see from the examples I have shared here, in some cases it makes browsing the internet a more pleasant and convenient experience.

This does not have to be an either-or scenario. You can still use the latest technological devices and protect your privacy, if you are aware of the tools that are available to help you do so. Much of the technology available from the open-source community has evolved considerably in recent years, making it easier to install and use. This gives you a choice about what level of privacy protection you want to introduce to your daily life, based on how you assess the threats to your privacy.

CHAPTER 10
Corporate

The solutions I talk about for corporations are really applicable for any organisation, large or small. I talk about government policy in the next chapter, but public sector organisations like the NHS in the UK really fall under corporate solutions as much as they do under government ones because they are run in the same way as a business, particularly when it comes to introducing and upgrading tech, even though they operate for the public good rather than for profit.

Why Do Businesses Need to Take This Seriously?

One of the strongest arguments for focusing on privacy and security online is reputational protection. What is seen to be protective and secure is increasingly being valued by prospective clients, as well as by the businesses themselves that are looking to buy both software and hardware. If, as a business, you do not take online security and protection seriously, you will find that you take a financial hit.

This is particularly true if you are a company that is developing and selling software, as we saw in 2020 with the SolarWinds hack, which resulted in the cyber security of many US government agencies being breached.

SOLARWINDS (OR WHY USING OPEN-SOURCE SOFTWARE IS A BETTER OPTION)

SolarWinds is a US software developer that has sold a great deal of software to both large businesses and government organisations, most notably within the United States. In 2020, its software was hacked, in what amounted to one of the most serious breaches of cyber security in US history.

It is believed that hackers infected the SolarWinds code as early as October 2019, which was then passed onto all of SolarWinds' client networks through a routine software update a few months later. SolarWinds estimated that 18,000 of its clients downloaded this backdoor into their software.[1] What is more astonishing is that the NSA, Homeland Security, the State Department, the US Treasury and many other US government agencies were among the SolarWinds clients that were hacked. So too were huge tech corporations like Microsoft and Cisco.

The hack was eventually picked up by a small cyber security company, FireEye, which notified SolarWinds and disclosed the hack to the press, which is how the whole story broke.[2]

The SolarWinds hack is a big wake-up call that becoming too reliant on one or two large software companies is dangerous, because it makes them an obvious target, but it also goes to highlight why open-source software is such a good option for corporations of any size.

Countermeasure: Open-Source Software

I explained in the previous chapter what open-source software is, so I will not repeat that here. However, what I will say to any business leader reading this is that open-source software offers your business the best chance of keeping your business, and the details of your clients, secure. When it comes to corporate security, you can be sure with open-source software that no backdoors are being built into the code. What is more, there is a global network of publicly minded geeks who are regularly checking the code and updating it.

Because it is open-source, you can also get your own in-house team of geeks to check the code and make sure there is nothing untoward in it. From a transparency perspective, it is second to none. This not only produces a sense of greater security within your organisation, but it also allows you to show your customers that you are taking their data and privacy seriously, and that you can be trusted. Open-source software therefore has multiple business advantages, beyond its security, in that it shows that your company is transparent, authentic and can be trusted.

You might be wondering why, if it is so secure and great, more companies do not use open-source software already. The answer (much as with us as individuals) is simply that they are not aware of its existence as an alternative to the closed proprietary software that is used by businesses the world over. This does, however, largely apply to companies operating in the West, because places like China will not allow open-source software to be widely available.

What Other Issues Can Arise from Using Closed Proprietary Software?

One of the major issues with closed proprietary software in the West is that businesses get charged significant amounts for licensing and then those systems need to be patched regularly. The problem occurs when an organisation has a system that is based on a very old version of this software, and the software company stops providing patches for it because it is obsolete. This puts the organisation that uses this old software in a very difficult position, where they may not be able to financially afford to upgrade to a more recent version and as a result are now no longer able to protect the data they are legally required to protect.

This therefore means that, as a business, you may be legally vulnerable to prosecution for not being able to protect your clients' data. I have already talked about the WannaCry ransomware attack, which affected many NHS trusts in the UK, but there have been many examples of hacks against health sector systems, including in the Netherlands, Finland and Norway. Protecting patients' medical records and their confidentiality is a legal requirement and, therefore, if health providers are unable to do so, they are leaving themselves open to prosecution or class action suits.

Why Is Open-Source the Better Option?

There are really two strands to the argument for open-source software, particularly in settings such as the health sector where budgets for technology are likely to be more constrained. The first is financial, in that even if you pay for open-source software, it is usually cheaper than what is offered by businesses selling closed proprietary systems. The second is that, even if you are unable to afford to upgrade your hardware, you at least have a fighting chance with open-source to maintain your protection, because the software itself is automatically upgradable and licence free.

This is particularly pertinent for the likes of government agencies, which tend to be more constrained when it comes to making large cash outlays. However, that is not to say that businesses cannot benefit just as much from making the switch to open-source software.

Using open-source software is also a way of future-proofing your organisation from what is to come, technology wise. Quantum computing, for example, is fast approaching and when that becomes a mainstream reality (probably by 2030, if not sooner), there are an estimated 20 billion devices in the world that will be out of date and, as a result, all the software that they use will be left completely wide open and vulnerable. By switching to open-source software before this happens, however, you will still at least have a fighting chance of maintaining a degree of security, because the software will continue to be updated.

The point is, you have to think about this from a long-term perspective. You cannot keep kicking the can down the road and telling yourself that you will look into an upgrade of your business's hardware "next year. . .". Even if upgrading your hardware is not an option, if you at least adopt open-source software you will be more likely to keep abreast of the evolving developments within the tech sector, including quantum computing (which I explore in much more detail in the final chapter).

Business Reputation: What Is Worth Fighting For?

In 2015, 14 people were killed in a terrorist attack in San Bernardino, California, when a married couple opened fire on a San Bernardino County Department of Public Health training event and Christmas party. The couple involved had used Whats-App to help plan the attack. The FBI took the company to court to try and get the backdoor key into this particular WhatsApp account to see what had been said and planned via the messages.

WhatsApp argued against the FBI, claiming that if they did this, then every other WhatsApp account would be left wide open. WhatsApp was known for being an encrypted messaging service and it went all out to protect its reputation. As it turned out, an Australian security company had already hacked into WhatsApp and to this specific account, but this did not stop WhatsApp from standing up to protect its reputation by claiming that if they agreed to the FBI's request, the security of all its other users would be threatened.[3] Personally I found it hard to stomach the pure cynicism of the legal case and what really leapt out to me at the time was that it appeared to show that WhatsApp was fighting for the privacy of its users and could therefore be trusted, when in actual fact it had already been hacked and as a result could not be trusted, regardless of the outcome of this legal case.

Secure Online Communication

Any business that wants to have secure communications around the world, whether that business operates in IT, health, finance or any other industry, needs to ensure that whatever communication system it is using can protect the information they are sending, discussing or writing online.

This has become an even more pressing issue following the Covid-19 pandemic, when remote working has become the norm for so many people. Many businesses are using closed proprietary software for their meetings, which is inherently not that secure. It seems that many people have just waved a white flag and accepted that they will have to use less secure options, but I think it will be very interesting to see what happens in the wake of the Russian invasion of Ukraine and how Western tech companies are responding to the unfolding crisis, as well as how they could potentially work to help protect Russian dissidents and those who are trying to push back against Russian government policy and the spread of misinformation (misleading information), disinformation (propaganda) and malinformation (true information that is taken out of context and the meaning warped) within Russia.

When it comes to secure online communication, many of the tech giants are, on the one hand, trying to make their products sound as secure as possible while also working to make them more secure. The issue is that these companies are often not doing enough. This is why the open-source community will often create their own versions of apps and software, because they can better guarantee their security. This is how the messaging app Signal came into being, and there are others popping up, such as Jitsi.

That is not to say the vast majority of people who work for these big tech corporations are not doing their job to the best of their ability. As we saw with the Snowden disclosure around the PRISM program, many of these companies (or certainly the vast majority of employees at these companies) were unaware of the backdoors that were built into their own programs. However, even when people are doing a good job to improve the security and encryption of the programs provided by large tech organisations, they are working against a stacked system. We already saw in Chapter 4 how the spies have kept backdoors and vulnerabilities to themselves. The issue with any closed proprietary system is that there simply are not enough people checking the code to make sure that these vulnerabilities are found and fixed quickly.

Open-source software works differently, because you have this global network of publicly minded geeks checking the code, identifying vulnerabilities and fixing them as quickly as possible.

What Can Big Corporations Do to Redress the Balance?

As I have said, these corporations have largely been preyed upon, much as we as individuals can be preyed upon. The challenge corporations face is that their entire business model is built on supplying closed proprietary software. That said, I am aware of some businesses in this sector that have invested a considerable amount of money in the application of open-source software, even though their business model is one that supports closed proprietary software. Some are looking forward to quantum computing and quantum cryptography and exploring how they can get ahead of, or at least keep up with, the curve of what is to come.

These producers of software are very sensitive to their reputations and are wary of taking reputational hits such as those that came out of the Vault 7 disclosures and the Shadow Brokers' hack of the NSA's cache of cyber weapons, as well as from corporate whistleblowers.

What About Businesses Buying This Tech?

If you are an organisation that uses this sort of software and typically buys off-the-shelf tech, my advice is to consider open-source solutions as an alternative to the closed proprietary software you likely currently use. This is one of the best ways to protect your business going forward.

For an example of how this can work in practice, we can look at the financial sector, which is where a lot of open-source, high-end tech has already been applied. Financial institutions have historically been tech innovators due to the necessity of being as secure as possible to protect their clients' money and therefore appear to pick up on tech security issues much more quickly than other industries.

Why Do So Few Businesses Consider Open-Source Software?

Aside from the fact that many businesses do not know of the existence of open-source software, one of the old arguments was that it was not as user-friendly as the available closed proprietary software. While that might have been true 15 years ago, it certainly is not the case now and, in my experience, open-source software is more intuitive and user-friendly than its closed proprietary counterparts.

For some large and long-standing organisations, another concern could be that all of their employees have used the same software and system for years, and therefore making a change will be painful in that sense. However, I would say that any company that is thinking like this needs to balance the perceived pain of making the shift to open-source software with the very real pain that would be inflicted on the business's bottom line and its clients should they be the victim of a hack as a result of their use of legacy closed proprietary systems.

The Business Case for Open-Source

There is a strong business case for pivoting to open-source software that is more secure and for taking more responsibility for data ethics. The first strand of this business case is that it is cheaper to use open-source software in the long run, once you have covered the upfront cost of bringing in an open-source specialist to design and set up the architecture for your company.

Even with open-source software, there will still be a maintenance fee if that system needs to be developed, evolved or even patched on occasion. However, where you save money is in not paying an annual licence fee like you do for the use of closed proprietary software. At a personal level, think about buying a laptop: it will come with a preinstalled proprietary software licence, which will likely have added a couple of hundred euros to the price of the computer. If you scale that cost up for a business, you can start to see how much that could save, even if you are paying open-source architects to remove proprietary software and install open-source software instead.

The second strand is in terms of the reputational benefit you can garner by taking this approach. You can explain to your clients that you are using open-source software because you respect their privacy, believe it is a more secure format, and it is easy to use. In addition, further down the line you can potentially benefit from improved security because open-source software, by its very nature, is not open to the backdoors and other vulnerabilities that I talked about in Part Two.

A third strand to this argument is that in protecting your organisation from cyber attacks by using open-source software, you are ensuring the business's continued survival. That might sound dramatic, but 60 per cent of small businesses go under within six months of suffering a data breach or cyber attack.[4]

Given that, in 2021, it cost businesses an average of $1.85 million to recover from a ransomware attack, that is not a big surprise. You also have to consider that only 57 per cent of businesses were successful at recovering their data using a backup, while even those that paid the demanded ransom only recovered 65 per cent of their data. For many small businesses, losing 35 per cent of their data would be crippling. Although the number of ransomware attacks being carried out has decreased since 2019, the amount of money being asked for with each attack has increased dramatically. Could your business pay the average ransom demand of $220,298, and would you want to pay out that much with no guarantee you would even get all of your data back?[5]

Ultimately, there is a better degree of protection for the software, and, by association, your business, as a result of using open-source software. While open-source developers (and the companies that work in this space) make their money by creating architecture for businesses, the technology itself is still free because it is not licenced, not to mention being more secure.

EVEN THE SPOOKS KNOW IT IS MORE SECURE . . .

One of the issues that sometimes gets forgotten in the debate about using open-source versus closed proprietary software is the matter of national security (although the SolarWinds hack has reignited this debate somewhat). The point is that if you have a nation state that is relying on off-the-shelf, closed proprietary software, it is laying its national security data wide open to backdooring, hacking, abuse or criminal activity via that closed proprietary software.

When I was working for MI5 in the 1990s, they were aware of this threat and as a result tried to develop an in-house system for the management of all of their information. As we know, knowledge is power, and if you are working in intelligence you need to be able to access this data very easily.

MI5 produced an in-house solution codenamed "Grant", but they just could not get it to work, so in the end they bought Microsoft Windows '95 off-the-shelf. This approach – and MI5 is far from the only nation state intelligence agency to have gone down this route – means that the agency is vulnerable, not only from hacks related to that particular software, but also in terms of using a piece of software that has been developed by a major corporation based in another country.

If you look at this another way, just think about the uproar surrounding Huawei being involved in the 5G rollout in the UK. Even before the invasion of Ukraine, would the UK government have asked a major Russian corporation to design and build software and architecture to be used across government agencies?

For corporations and public sector organisations, particularly across Europe, I think there is a reputational win-win to being seen as trying to peel away from this dependency on American corporations, as well as from working to develop more secure, open-source systems.

Do Not Forget About Physical Security

For all that I am talking about data, the internet, and our online security and privacy, in a corporate setting it is important not to forget that if you have a physical office somewhere, it is very easy for your software to be tampered with. All it takes is one person to bring in malware on a USB drive, for instance, and it can infect your whole network. Depending on how firewalled all the different segments of a multinational corporation are, this could then spread from country to country, taking out your entire business infrastructure.

Open-source software is, once again, the obvious countermeasure for this. Not only is it more secure, it is also more sustainable. Once you have open-source software, it will automatically be updated to repair any vulnerabilities, without the need for your organisation to buy outdated patches for the software you have the licence to. This means you are going to be far more effectively protected.

It seems that cyber security, in a budgetary sense, is an after-thought for many businesses. They do not regard updating their computers, systems and licences with the same sense of urgency or importance as other expenses within the business. However, because they do not spend the money to do this properly, they leave themselves more vulnerable to a costly attack. When Sony was infamously hacked in 2014, cyber security researchers criticised the company for its poor data security and questioned how much of a priority cyber security was for the organisation.[6]

The Business Opportunities in Open-Source and Better Data Security

For all organisations, regardless of industry, data privacy has become part of the "products" they sell. In the world we live in, that is unavoidable. Over the years we have seen a great deal of reputational whitewashing from companies (just look at the WhatsApp example I shared earlier in this chapter). It is one thing for a company to say that it delivers end-to-end encryption of messages, for example, and it is entirely another for it to actually deliver on this promise.

If, however, your business does place a great deal of importance on data security and puts this at the centre of what you do, it is important to create that narrative around your activities and to be authentic in doing so. It is important to help people understand the bigger picture, in that this is not only about software, but also hardware and firmware, and to explain why you are taking this stance on data security.

Consumers are increasingly aware of the level of predatory behavior they face online and they are increasingly aware of the lies they have been told by other organisations that claim to be secure or to provide privacy. Therefore, being transparent about how data is stored and used within your business, as well as how you are protecting consumers against potential data security breaches, will attract more consumers to you.

I am talking about being completely honest, rather than sweeping what you are doing under the carpet. Consumers not only appreciate that honesty, but are seeking it out more and more. Being a completely open book with your customers is not only more important than ever before, but also more valued. As a business, this should matter to you as well, because you can be named and shamed on social media and elsewhere online very easily. The reputational damage that can come from being perceived as not being honest and transparent can be substantial. Once broken, consumer trust is almost impossible to repair. It is far better to put your business in control of the situation by genuinely working to protect your customers and their data, and communicating that honestly and openly.

Countermeasure: Blockchain Technology

Blockchain is a way of data stamping the packets of data that get sent all over the internet for authenticity. That means every time the data passes through any sort of change or transaction, it is stamped. Blockchain was originally developed for the cryptocurrency Bitcoin and, unfortunately, due to this association with the fintech industry and some of the sharks in that sector who have not used it responsibly, it has suffered some reputational damage. However, there are other applications for this technology that can very much be used for the good of all of us.

For example, when it comes to your medical records, using Blockchain you can see where your data has been, who has access to it, and who has accessed it because each time any of these packets are accessed they are stamped. This also means that you are able to see if anyone has tampered with the data. It can also potentially be used for secure electronic voting in a fair democratic election, which is something that a number of us have been exploring, especially in the wake of all the horror stories we have heard in recent years about how easy electronic voting machines can be to hack.

It is direct recording electronic (DRE) machines that are the most vulnerable, especially as some do not even provide a paper trail of the votes cast. There is evidence that these can be physically tampered with, that hackers are able to design multiple-use election cards for DRE machines, thereby enabling people to cast multiple votes, and that DRE machines can be subjected to targeted attacks using public Wi-Fi networks.[7] Equally concerning is the ease with which websites that report the results of elections can be hacked. While tampering with the results being displayed will not ultimately affect the ballots themselves, it is enough to sow the seeds that an election result is not legitimate. At Def Con in Las Vegas in 2018, it took one child just 10 minutes to hack a website and alter the results displayed as part of a controlled hacking exercise.[8] This does not bode well for the future of fair and free democratic elections.

Blockchain technology therefore offers potential advantages when it comes to protecting one's data as it passes down the fibre optic from end to end. It can also work with Tor technology, even though you are jumping around from node to node, because you can still see who has accessed the data, who has tampered with it or who has validated it and used it for good. From a societal evolution perspective in terms of future-proofing our data, or in fact any data that moves over the internet (pre-quantum computing at least), this is the best technology we have available to us.

In Chapter 7 I mentioned a project we are undertaking at the World Ethical Data Foundation that is working to future-proof information, specifically at this time around human rights, using Blockchain technology for distributed storage. This means that the information will be stored all over the world, because even if a country is safe now, in years to come it might not be, so by using Blockchain to data stamp everything to verify its authenticity we can store data anywhere and in doing so future-proof it. This particular project is also linguistically neutral, as I mentioned earlier, which we are achieving by using AI and data learning, while the end user interface is all open-source.

Blockchain truly is revolutionary technology, but it has not been exploited effectively yet because the money has not been there to make it valuable. However, it presents some fantastic opportunities for the greater good and has many potential business applications outside the world of fintech. As I write this we are working on a number of projects at the World Ethical Data Foundation around how businesses can adopt Blockchain technology and exploring new use cases. Watch this space!

CHAPTER 11
Government

My area of expertise is within government policy and over the years I have given evidence to various governments and government organisations to highlight the potential issues policymakers and lawmakers need to consider in relation to data protection, online privacy and our online rights, as well as to provide constructive suggestions that can be incorporated into national or supranational laws.[1]

For example, in 2014 myself and a few other experts in this space testified at a European Parliament special hearing about many of the issues that came out of the Snowden disclosures. Among those who spoke at that hearing was Tom Drake, an NSA whistleblower who, like Bill Binney, went public with his concerns about the NSA's activities following the failure of the Trailblazer project. One of the steps the EU took following this hearing and others, for example, was to formulate a directive in the EU to protect whistleblowers. In 2022, the European Commission also put forward a Declaration on Digital Rights and Principles,[2] which at the time of this writing is expected to be ratified within a matter of months.

Having a Declaration on Digital Rights and Principles is all well and good, but the question remains about how this can be meaningfully implemented given what we know about the power the spies and corporations have. This is especially the case when you consider Article 10.2 of the European Declaration of Human Rights, which provides a blanket exemption around national security when it comes to freedom of expression. This means that every country in the EU still has primacy over the laws they choose to adopt when it comes to what spies or the military can do. Essentially it is a "get out of jail free" card for intelligence agencies.

The other question is what the implications will be for companies. If we look at the GDPR directive that the EU introduced in 2016, we see that not only did it cause companies quite a lot of pain in trying to implement it, but that there is still ongoing friction as a result of GDPR, particularly between US corporations and the EU.

At the time of this writing, full details of what is contained in this Declaration on Digital Rights and Principles have yet to be shared; a major concern with legislation of this nature is that it might sound good and talk about respecting people's rights online and ensuring that we have access to safe technologies, but how will any breaches be enforced? Both the EU and UK governments have struggled for years to effectively enforce even the collection of tax from these big tech corporations, so if they cannot manage that, how are they going to enforce our human rights and privacy in a digital world?

There is also a great deal of talk about ensuring that people have individual choice when it comes to the technology they use, but if there is no talk of any options beyond the closed proprietary software providers, then people are not being given a meaningful choice because they are not being made aware that there are alternative options available to them to better protect themselves or their businesses.

One of the major challenges governments the world over face when attempting to introduce meaningful legislation to protect our privacy and human rights in a digital world is that elected MPs or MEPs are not experts in this field. They will listen in good faith to one perspective on the issue, such as that put forward by the various digital rights organisations, and they are then lobbied hard from the other perspective, which tends to be in the interest of big, multinational corporations.

The issues being debated now and the potential enforcement of any legislation enacted in the coming years are also not looking far forward enough. With AI and quantum computing hovering over the horizon, our policymakers and lawmakers need to have access to the right kind of expertise and advice to formulate a roadmap that will account for this, which at present simply is not happening. For example, the EU has stated that its Declaration on Digital Rights and Principles will form part of a roadmap to take us up to 2030; however, we will certainly be enmeshed in quantum computing by then, which will be an incredible game changer.

Government Action Has Become Imperative

As a result of the Covid-19 pandemic, we are all living much of our lives online and we no longer have a choice about that. It is not our choice whether to set up a Zoom account (or not), because we need one for work. We no longer have a choice about whether to have an Instagram or Facebook profile, because they are expected as part of our professional lives. Because we have lost our right to choose whether or not to engage in this online world, it is more important than ever for governments to recognise our human rights in every sphere. As I explained in Chapter 1, you cannot separate digital rights from human rights – they are one and the same. We are humans and we express ourselves in different ways; doing so through tech is just another expression of our humanity and our rights need to be protected in the online realm, as they do everywhere else.

Reaching a Fork in the Road

The UK government's policies in this area will be particularly interesting to watch in the coming months and years. As part of its post-Brexit activities, the UK government is going through a series of consultations, including surrounding issues having to do with privacy and digital rights. Following Brexit, the UK is now outside of EU legislation such as GDPR (although it has retained it for the time being, the government has stated that it will keep the framework under review), which means the country has an opportunity to reinvent itself in one of two ways.

The first is to allow the UK to become something of a "digital free-for-all" offshore of the EU, and the second is to think outside the box and consider how to become a beacon of light in terms of our privacy and human rights as we live our lives increasingly online. Post-Brexit, the loss of EU GDPR protections means that the UK is at a real tipping point in this sense, essentially where it has the opportunity to either descend into the digital sewer or to become a beacon of light for the rest of the world to follow in terms of how to manage and protect our human rights in a digital world. The World Ethical Data Foundation, with which I work, has been in discussions with the UK Cabinet Office to instigate a series of deep dives in areas such as quantum computing, privacy, and AI ethics.

Of course, the UK already has one of the least accountable and most legally protected intelligence agencies in the world (due to the Investigatory Powers Act 2016, which I talked about in Chapter 1); however, there are people within the UK government who recognise the corporate and reputational benefits of stronger protection for our privacy and rights online, as well as the ethical and moral benefits of being seen to do the right thing as a national economy.

These competing interests at a national level will impact privacy for individuals, companies and government institutions, as well as the reputation of the nation within the international community. Whether the UK ends up in the sewer or shining a light for the world to follow remains to be seen.

A Question of Knowledge

One of the big challenges that governments, certainly in the West, face is the way in which policy is formulated. For example, within the UK civil servants are tasked with producing policy papers, which are then presented to specialist committees. In a field such as health, some of the MPs sitting on that committee might previously have been doctors; or when it comes to defence, some of the MPs on the committee will likely have served in the armed forces. However, when it comes to technology the number of experts in this field within the MP demographic is generally quite low. This means that those sitting on the specialist committee to review digital policy may not have a significant level of knowledge themselves.

Given how quickly tech is developing, this only adds to the challenge of finding politicians who have the depth of knowledge required. As a result, MPs will often rely on their staff, but civil servants often are not technologists either. To bridge this gap in knowledge, what tends to happen is that this work of producing policy papers is outsourced to large consultancy organisations. These consultancies are very good at what they do, but they also tend to have tunnel vision and do not always see the whole landscape. This can therefore mean that the MPs who are responsible for dealing with these policies and formulating new laws in this area do not necessarily receive a broad spectrum of advice.

The acceleration of the development and use of technology, and the degree to which it underpins all of our lives, has made it very difficult for policies and laws to catch up. In fact, governments have been playing catch-up (and are still chasing technology) in the tech arena for at least 30 years. What I, and many others who work in this field, have been stressing for many years is that technology is not only vital in terms of business and finance, but also for our democracy and society. It underpins everything we do in the modern world.

For too long, certainly in the UK at least, the digital world has been considered something of an afterthought. At the time of this writing, policy in this area is overseen by the Department for Digital, Culture, Media and Sport, which indicates that it is not considered a particularly high priority. There is certainly a strong argument for governments having a department solely dedicated to regulating, monitoring and auditing the technology, data and digital sector, although that sadly is not happening.

I have been talking about the issues I have laid out in this book for over a decade, but it is only in the past five years that cyber security companies have approached me and asked me to speak to their teams about the depth and breadth of the threats we face in a digital world. That speaks volumes in itself and we should all be concerned that these cutting-edge cyber security and defence companies have only recently shown an interest in learning more about this area. These are the organisations that provide consultancy services to governments and that win the contracts to build the infrastructure our government agencies use. They therefore need to be ahead of the curve – but in reality they are often falling behind it.

How Can Governments Redress the Balance?

A great starting point for any government would be to set up a department that is solely dedicated to technology, digital and data, rather than lumping it in with other policy areas. This area has become something that is fundamental to the functioning of our society and that impacts our lives in multiple ways on a daily basis as individuals. Rather than focusing all the attention on cyber attacks or defence, and seeing the experts we do have within government funnelled off into roles that focus on this capacity in a militaristic sense, we instead need to look at cyber security from a societal perspective.

Governments need to look at how they can protect both their citizens' and businesses' privacy, needs and rights in a digital realm. Of course, there is an element of this that needs to be militarised, but my point is simply that this should not be the sole focus. As things currently stand, we have a situation where we, as citizens, are being asked to trust our governments to protect us and whenever there is a question over how government agencies are protecting our needs and rights, the response is, "We cannot tell you, it is national security".

We need to have more open discussions and consultations about cyber security and how we can all be safer in an online world. There needs to be more plurality and engagement with those working on the civil side of technology, digital and data, as well as continuing to involve those working on the corporate side of these fields. We need to encourage governments to listen to other voices in this debate, including those who are often marginalised, like members of the hacking community. All too often hackers are broadly branded as criminals, when there are many white-hat hackers who work for the common good. Only by bringing these diverse perspectives and broader expertise into discussions can we hope to develop a good, ethically based, forward-looking infrastructure. If the UK chooses to do this, it will be well on the way to becoming a beacon that others can follow.

Education Is the Key

Once again, education is a cornerstone to improving the digital landscape both in the UK and elsewhere. Many people working in government organisations are simply unaware that there are very viable and cost-effective software options available beyond the closed proprietary systems they currently use, for instance. This means it is highly likely that the people making decisions about technology in these settings will simply go for the easy option and renew existing licences without looking at what else is available.

There is often a false sense of security that comes with using these well-known providers, in that people believe that vulnerabilities and issues will be automatically patched. However, as we know from earlier chapters, there are vulnerabilities these companies are simply unaware of and if those find their way into criminal hands, they can and will be exploited.

Education of our elected officials and politicians is therefore a key part of this puzzle, because it will help them understand the wide range of options open to them and could set some governments off on a different path.

Open-source software is one solution that I keep coming back to because it provides so many benefits, from being more cost-effective than the closed proprietary systems to being more secure. There is another benefit to open-source in an educational setting, too, in that introducing open-source software to schools will start to prepare the next generation for life in a digital world.

There have been stories about some schools requiring students to have Facebook accounts simply so they can interact with their teachers online. This sets an entire young generation on a path of thinking that all interaction through the internet is via a corporation like Facebook, which of course is not true.

Countermeasure: Promote Tech Experimentation in Educational Settings

There are a number of ways in which to encourage the next generations to experiment and better understand the technology that will shape their lives. Raspberry Pi is one such example. This is a small and affordable computer that can be configured in various ways and that encourages you to learn to code. If you put this in the hands of children and teenagers, they will start to play and experiment with technology in a way that teaches them about how it works at a meaningful level.

One Dutch charity sent repurposed laptops to Caribbean schools and these laptops only had open-source operating systems and programs installed. Not only is this an affordable way to put technology in the hands of those who need it, but it also means the kids who received the laptops could use them and learn to code. This project, and others like it, are building up the educational knowledge base for the next generation of citizens, rather than making them dependent on a particular style of software and perpetuating the myth that there are limited avenues available to use the internet.

When you build up this wealth of knowledge among your citizens, both your country and the businesses based there have the potential to follow a range of directions. This is likely where the next big technological breakthroughs will happen.

Using open-source software in an educational setting also has a social justice aspect, in that it makes technology more easily and affordably available to children and families who otherwise would not be able to access it. India has been using this model very effectively for many years, where it strips old computers back, installs open-source software on them and resells them. This has multiple benefits, first and foremost making access to computers much more affordable to those who cannot afford to buy new devices. By using open-source software, copyright issues are avoided, and this is allowing citizens in India to develop their knowledge of coding.

These kinds of schemes can also be implemented in countries in the West, where children whose families are not able to afford their own laptops or computers can access older, repurposed devices at a lower cost. The key is to make sure that everyone who receives a computer has a choice about what software is installed on it. If you simply give out computers with closed proprietary software on them, you limit children's thinking around technology. If, however, you give them a device that they can adapt and tinker with, you give them far more freedom and accelerate their ability to learn about coding and other aspects of technology.

Most of the geeks and hackers that I have met grew up using closed proprietary software and only discovered alternative options like open-source software when they started learning to code and found an online community of like-minded people. It took effort on their part to uncover these alternative options, but it should not be so difficult for all of us to learn about the alternatives that are available. By introducing children and teenagers to open-source software at school, you are showing them that there is a choice and that there are a wealth of possibilities out there. Just giving young people devices and encouraging them to use whatever closed proprietary software is installed on them is not educating them; it is just another way of enforcing consumerism.

Failing to introduce the next generation to all the options is potentially damaging not only to a country's knowledge base, but also to its economy and society. If you only raise young people on closed proprietary software, it is a bit like teaching them to swim in a pool in a landlocked location and never telling them about or showing them the ocean. One day they might make it to the coast, but how many will go looking for it if they do not even know that it is there? And will the ones who do find it be able to swim in the waves?

We, as a society, have to teach young generations of digital natives how to empower themselves with technology, rather than becoming slaves to it. By giving them all the options, they have a much greater chance of being able to take control of their lives online and harnessing their passion and desire to effect change.

Creating a Cultural Shift

There are several places where we can look to create a cultural shift within our societies. The first would be with the capitalist system itself. I would argue that the United States is at the hard end of capitalism and it is where we have seen the most aggressive corporate growth strategies in terms of the big firms buying up smaller ones, particularly in Silicon Valley. There is another option within the capitalist system, however, which is an approach that focuses more on a social democracy model.

The Scandinavian countries and Germany are examples of where this can work. In fact, in Germany they even have statutes in place to encourage and protect small and medium-sized companies, rather than allowing these organisations to be hoovered up like krill by some giant corporate whale. Creating an environment where small and medium-sized businesses can flourish rather than being threatened all the time provides cohesion and strength to society. It is the softer end of capitalism and one that the tech industry could benefit from adopting.

Another area where we need to shift our thinking is in relation to the technological infrastructure itself. At the time of the first dot-com bubble in the 1990s, there was a great deal of concern in Europe about an overdependence on the US-built infrastructure that ran the internet. In July 2001, the EU published a report about the need to move away from this dependence on American fibre optics and telecoms. The intention was to start building Europe's own infrastructure and in doing so to move away from the potential predations of US corporations. However, then 9/11 happened and a lot of this got shelved.

The EU is not the only part of the world that has expressed concern about relying on infrastructure provided by corporations from outside its nations. Following the Snowden disclosures, Brazil also announced that it was intending to build its own transatlantic fibre optic cables, although this project is another one that has struggled to gain traction due to the country's complex political situation.

Much of this argument comes back to the concept of net neutrality and whether this is at risk by allowing corporations to control so much of the world's digital infrastructure. Net neutrality, as I explained in Chapter 5, is the idea that everyone, whether a business, government or individual, should have the same degree of access to the internet. There is an ongoing battle about this, where some believe governments or corporations that want to pay for better connectivity and higher speeds should be able to do so, while others (myself included) want to ensure that everyone has access to reliable, high-speed internet.

That said, even when the concept of net neutrality is enshrined in law, it does not always happen in practice. You can see this in action in Guernsey, which is where I grew up, so this is a very personal example for me. Despite having a wonderful high-speed fibre optic connection to the island, many people there still have unreliable internet connections. The reason is that Guernsey has become an offshore tax haven that is infested with bankers, and 95 per cent of the island's connection is used by the finance professionals and lawyers based there, leaving very little bandwidth for the rest of the people who live there.

There are also many parts of the world that do not have the same level of technological infrastructure that we do in the West. Parts of Africa, Asia and Latin America all struggle to access reliable and fast internet connections. It will more than likely be corporations, rather than national governments, that introduce updated infrastructure to these regions of the world, which leaves governments with another challenge in terms of how they can rein in the particularly predatory practices of large corporations.

As I have already mentioned in this chapter, education is key for these cultural shifts to take place. We have to make sure that everyone knows about their options when it comes to the devices and software they use, and governments have an important role to play in raising that general level of awareness in society.

But we also need to ensure that our policymakers, lawmakers and politicians are better educated about all the different tech options available, because if they do not have this knowledge, how can we expect them to even begin to think about building an infrastructure, education network or companies that make hardware that can protect the rights and privacy of their citizens and the intellectual property of corporations?

The Driving Forces in Cultural Change

Generally speaking, any large-scale cultural shift like the ones I have been discussing here have to come from the bottom up, and that will only begin once people realise there are other options and that the way we have been living to date is not the only way to live our lives. The crux of this argument, however, is that if people cannot access the information they need to bring societies to this cultural tipping point, there is a blockage to any cultural shift taking place. This is one of the main issues we face with the tech we use, because many people are not aware of what all the options are and they cannot always access the information they need to realise there is a big social movement out there pushing for this change.

We looked at the control of information in the media in Chapter 7, and it is important not to exclude social media from this conversation. It might sound fanciful, but we have seen on many occasions in recent years the power that social media companies have to restrict the flow of information when they want to – if even a sitting US president can be banned from Twitter, what does that tell us? In addition, the algorithms these companies use create online echo chambers, which further restrict the diversity of information and perspectives we see.

Essentially this indicates that the ideas or movements that such corporations approve of will be allowed to flourish and those they deem unacceptable will not. If we, as citizens, cannot access all the information and data we need to make our own decisions about whether or not we need or want to be active about these issues, it is very difficult to affect change at a societal level.

The argument always comes back to access to information. The internet is the new version of the Gutenberg Press and, as we know from history, dissemination of information is key to mobilising any large-scale movement for societal change. If you look back through history, you can see how the pace of cultural change accelerates as it becomes easier to get information into the hands of the people. For example, the UK was the first country to abolish slavery, and it took Wilberforce around 30 years to achieve that goal. Through the actions of some very brave women, universal suffrage was achieved in about 20 years. Then if you look at the Stonewall movement for the legalisation of homosexuality, that happened even more quickly.

All of these cultural shifts happened pre-internet, and now technology gives us the opportunity to create these cultural upswells even more effectively by more quickly normalising something that once seemed different. However, this is where control of the internet and technology becomes crucial, because if access is being blocked or restricted in some way, it makes it much more challenging to create the desire for a cultural shift within populations from the ground up.

Of course, it is also important for a desire for change to come from the top down, which is where governments come into play. To make legalistic and government-level changes, those in government need to be able to see all of the perspectives around an issue and fully understand the potential pain a situation is causing their citizens in order to encourage them to make changes on a legal level.

This comes back to having a plurality of perspectives, coming not only from lobbyists but also from specialists and those in civil society. All of this helps to build a groundswell of people who understand a situation from a cultural perspective.

An example of where this has worked in a different policy area is in relation to ending the war on drugs. One of the roles I had over the years was with what is now the Law Enforcement Action Partnership, which provides a platform where people from every level of society can add and amplify their voices around this issue, and which national governments and supranational organisations like the EU and UN engage with. As a result of this work, we are seeing a cultural shift in perspective around the war on drugs and when this is positioned correctly with the right organisations, it is able to amplify that argument and this in turn can nudge the cultural tanker in a new direction. In this instance, this is in the direction of a more care-focused environment rather than one that is focused on penalisation and prosecution.

We need to see this kind of engagement and discussion around our rights and privacy in the online world. Governments have a crucial role to play, not only through setting policy but also by supporting the education of their citizens in the tech arena. It is only when all of us work together for the greater good that we are able to steer the cultural tanker toward a utopian scenario and away from a dystopian one.

CHAPTER 12
Utopia or Dystopia?

What kind of world do you want to be living in 30 years from now? As I write this book we are on the brink of significant technological changes that could push our society closer to the kind of online utopia envisaged by the early internet pioneers or that could push our society toward one that more closely resembles George Orwell's *1984* than any of us would want.

We have reached a pivotal moment in society, one in which we have the ability to make all of our lives better. However, if we let technology go too far and too fast without reflecting on the issues I have outlined in these pages, we are going to lose that chance to build a better and more equitable future, especially as quantum computing accelerates and becomes more mainstream. It is time to make a choice.

Utopia . . .

As you close your door behind you, you smile. In your hands is the parcel you have been waiting for. It is your new home hub and you cannot wait to start customising its code to make it work for you and your family. You sit in your living room and check the news; the government has just announced more funding to support coding lessons in primary schools.

You are itching to get started with customising your new piece of tech, but first you jump into the metaverse to get some tips from some of your friends. As you strike up a conversation with Ash, you immediately notice that their avatar has gone from female to male – you smile to yourself, knowing the pleasure they probably took in designing their new avatar to perfection. Conversation quickly turns to your new home hub and how best to program it.

As the weeks go by, you repeatedly tinker with the code, dipping in and out of various groups in the metaverse to get tips on how best to program your hub. You are also looking into attending a protest at the weekend, with many others who are concerned that the government is not going far enough to tackle the climate crisis. There are lively discussions about the protest itself, where and when to meet and what tactics to use to make sure your voices are heard . . .

Dystopia . . .

As you close your door behind you, you know you should feel safer, but you feel uneasy, even in your own home. You sit in your living room and check the news; the government is pushing through new laws that will crack down on those protesting about the environmental and climate catastrophe that is unfolding around the world. As you read about the details of this new legislation, your stomach does a backflip. Anyone who has ever attended a climate crisis protest will be monitored to ensure they are not "environmental extremists". You know you attended various climate marches in the early 2020s; what does that mean for you now?

Looking around your home, you are suddenly aware of all the devices that the authorities could be listening in on and monitoring. Your avatar in the metaverse has always aligned itself with organisations and businesses fighting for better environmental policies. You log in, only to find that several groups you were a member of have had their meeting places closed. You do not know who to reach out to, or how to do it. Every option in the digital world suddenly feels very unsafe. You do not know who might be listening.

As the days and weeks pass, you become increasingly paranoid. You are just waiting for the authorities to swoop in. You have done your best to erase some of your avatar's past data in the metaverse, but you are not convinced of how effective you have been. Nowhere feels private or safe any more. Even during a walk in the park you are suddenly aware of the CCTV cameras dotted around, which feel as though they are tracking your every move . . .

Which of these worlds do you want to live in? After my experiences in the 1990s, I can certainly identify with the dystopian vision of the future and that is not a world I want to go back to. You have the power to decide; we all do. We also have the power to push our politicians into action to make sure that our future looks more like utopia and less like dystopia. If we can begin to tip governments into thinking and talking about these issues, challenges and evolutions, and perhaps reining in some of the baser corporate instincts we see in the world, we have the ability to create a world where there is a free flow of information, knowledge and freedom of expression.

The Move Toward the Metaverse

As I write this, the metaverse is a hot topic of discussion and, whether we end up with a digital reality run by Meta or another organisation, this is certainly the direction in which we are travelling. This, of course, raises questions about who is going to own the data farms that host "us", who is going to own the software and who is going to have access to what we are in terms of intellectual property issues when we are living online in this virtual world. It is certainly a potential future threat if we get it wrong.

This does not only apply to those of us living in the West. China is also going down this path at an accelerated pace and they have a very different political and cultural environment. There has been a great deal of talk about China's digital social credit system and how that will evolve in the coming years, for instance. We already know that those who do not accept such a system in a country like China will pay a far heavier price for their dissent than those of us living in the West.

Our direction of travel risks sounding like something out of a sci-fi novel – Iain M. Banks described what can only be described as a geeks' utopia in his Culture novels. It is a universe in which you can upload your consciousness, have a body printed and download your consciousness into that body. You can change your sex, even your species. Everything is possible. However, it is not too much of a stretch to say that society, certainly in the West, is taking its first tentative steps in that direction with the development of the likes of the metaverse and the vast range of technical projects a multitude of tech billionaires in the West are developing as we speak.

This kind of world has the potential to become dystopian or to become utopian. If these virtual worlds are owned by corporations, which can therefore then control us, who we are and how we are living in their particular universe, we are moving toward dystopian territory. If, on the other hand, we can create these virtual universes using more distributed technology (such as the blockchain method of storage) and more open-source technology, we are heading in a more utopian direction.

This all comes back to making sure that all of us have the knowledge and skills to manage our lives online and to protect our human rights as we express ourselves in an increasingly virtual world. That means each of us having the capability to program and code to help ensure our personal privacy, and also to simply be mindful of these potential issues as we explore these new digital realms.

Companies that are thinking along the lines of how they can support people in becoming more educated, and how they can protect people's privacy and rights online, will increasingly be appreciated and recognised for taking that stance. Similarly, governments have a genuine opportunity to steer us in one direction, as I discussed in relation to the UK in the previous chapter. They are at a fork in the road and can choose whether they become a sewer of data or a beacon of light leading the way in our digital transition.

In an ideal world, we as individuals will all have the ability to take control of our own identities, how we want to be perceived and how we want to interact as we move into that digital world.

Quantum Computing: A Game Changer

Quantum computing is a true game changer in the way in which computers work and the speed at which they can compute. They are going to be so damn fast! All the research suggests that what would take a current supercomputer a week to do could be carried out in just a second by a quantum computer. This presents a very real threat to all of our existing security models in the online and digital world, even those that are very strong and that have been working well for 25 or so years. Quantum computing will simply blow all of them out of the water.

We have to ask where that will leave businesses and us, as individuals, who are living our lives online for both work and pleasure. The arrival of quantum computing will mean the end of privacy as we know it. It truly is a giant leap forward and, a bit like when the internet was invented, it is going to change the world. Suddenly, all the steps you have taken to protect yourself in the online environment as it is now will be nullified.

As well as being aware of this on an individual level, it is essential that governments start to think about how this will affect our privacy and our rights. Policymakers need to start thinking along these lines and how this technology will fit within either existing policies or those that they are currently working on to try to provide better protection of our basic human rights in our increasingly digital world.

The worry is that if this technology starts being used, without any forward-thinking legislation to protect people online, we will end up in another Wild West scenario where there is a battle between businesses looking to develop technology for the good of society and those looking to develop technology solely with a focus on maximising profits.

The general consensus is that we will have quantum computing by 2030, potentially sooner. With all the strides being taken in this space commercially, it is fair to assume that quantum computing already exists on some level in a military and security context, because these sectors are usually about a decade ahead of the corporate market.

SEEING THE FUTURE

My father was a newspaper editor and I remember him attending an editor's conference in the UK in 1997, where they looked at evolving tech. When he came back from that conference, he told me that there were contact lenses that could connect you to the internet. He was never terribly specific about who had been showing this particular innovation, but this was likely military-grade tech at the time. Fast-forward to the late noughties and Google introduced the first commercial application of this tech with Google Glass.

This simply demonstrates that there can be incredibly advanced technology available in some settings that most of us are unaware of because it has not become mainstream yet.

To give ourselves a fighting chance of being able to control our virtual lives as well as our physical lives, we need to focus on education. Learning to code and being able to understand what is happening under the hood of the technology we use is a vital aspect of this and one that is only achievable if there is a much broader shift toward open-source software and technology. For younger generations who are going to be growing up and living in a world with quantum computing, it is only fair to teach them the skills they need, like coding, to be able to develop their own systems and defences.

This connects closely to what I discussed in the previous chapter, whereby it is only through education that we can hope to create the democratic and cultural will to affect change for the benefit of society as a whole. By providing a really good education for the generations following us, when this technology becomes mainstream they will be in a position to understand and deal with it in a manner that protects their human rights and privacy, rather than blindly using and consuming it. We have to end our air of complacency about the role tech plays in our lives if we are to create a utopian, rather than a dystopian, future.

The fact that large corporations, such as Meta, are exploring options like the metaverse indicates that they already know that their current business model will be superseded. In many ways, you could say that the big tech industry as it currently stands is analogous with the old music and film business from the twentieth century, which initially fought against developments like online streaming only to be forced to embrace them and change the way they operate in the process. A similar shift is coming in the sphere of the tech giants.

It Is Time for Governments to Wake Up

If corporations are already seeing the writing on the wall for their business model, it is time that governments also wake up to the future we are moving toward. The challenge, particularly in the West with its four-year election cycles, is that politicians very much take a "here today, gone tomorrow" attitude.

The majority of governments in the West have civil servants, who are supposed to be politically neutral and to ensure that the country is run regardless of who is in power. In recent decades, however, we have seen an increasing number of politicians introducing "special advisers" to their teams, who are typically brought in (and then leave) with a specific minister. This is eroding the longevity of the knowledge held within the civil service and this means that Western governments have become overly reliant on consultants and corporations when making decisions. This has contributed to issues of tunnel vision, which I explored in the previous chapter.

Through my work as an activist, I have met many passionate people who are concerned about the issues I have outlined in this book and are able to join the dots to connect them (a skill that is sadly lacking in the political establishment in general); however, they often do not have a voice. They are not being given a seat at the table to share their very valid experience and perspectives.

My other concern is that schoolchildren, as well as the rest of the population, are not being presented with all the options out there when it comes to tech and managing our lives in this ever-more-digital world. Our lives are largely lived online through coercion rather than choice, thanks to the Covid-19 pandemic, and the problems I have highlighted in relation to our individual rights and privacy are only going to be exacerbated if we do not educate people about all of their options from the ground up.

Furthermore, corporations seem to have a whip hand when it comes to governments and international organisations. Lobbyists seem to be able to steer and change government policies that have been made with the best of intentions (remember the example I shared in Chapter 3 about how the Dutch government mandated that all public bodies should use open-source software where that was an option, only for the law to be repealed as a result of corporate pressure?).

Education: The Binding Thread

Education truly is the thread that binds everything together and that will lead us toward a utopian rather than dystopian society. First, this thread is about education around how you, as an individual, can use technology. Second, it is about education around how our policymakers can protect our use of technology. And third, it is about protecting our perception of what is going on in the world and ensuring we are able to educate ourselves and access diverse perspectives, either via the mainstream or online media.

The Time to Act Is Now

We have to act now, though, before it is too late. If we fail to educate ourselves and wider society, we will all become too boxed into this digital world in which we live and we will lose our basic freedoms. There are opportunities for governments to make the right choices. There are opportunities for emerging technologies to be used correctly if there is education and awareness of their potential. There are opportunities for corporations to demonstrate social responsibility and to thrive on the back of that.

We stand at that fork in the road. We can make a choice for a society that is focused on protecting our human rights, freedom of expression and our privacy; or we can choose a society that sells our data and erodes our human rights and privacy in the name of profit and so-called national security.

We have the power to tip the balance one way or the other. Ask yourself what kind of world you want your children (or young people you care about) to live in. The generations that come after us will need to live with the consequences of the decisions we make now. What can you do to help educate them about technology and how it can empower rather than control them? Where do we want to put our trust and who do we feel we can trust? Can we trust our governments, the corporations and even ourselves in terms of the choices we make?

By spreading knowledge and education about technology and the myriad issues I have highlighted in this book, we can empower ourselves and others and create a world where our digital and physical lives intersect in spaces that are private, free and protected.

NOTES

Chapter 1: Our Human Rights Online

1. Datt, A., 'The impact of the National Security Law on Media and Internet Freedom in Hong Kong', *Freedom House,* 8 September 2021, available at: https://freedomhouse.org/article/impact-national-security-law-media-and-internet-freedom-hong-kong.

2. 'Saudi Arabia: 10 reasons why women flee', *Human Rights Watch,* 30 January 2019, available at: https://www.hrw.org/news/2019/01/30/saudi-arabia-10-reasons-why-women-flee.

3. 'Saudi Arabia', Human Dignity Trust, available at: https://www.humandignitytrust.org/country-profile/saudi-arabia/.

4. 'Internet shutdowns in 2021: the return of digital authoritarianism', *Access Now,* 28 April 2022, available at: https://www.accessnow.org/internet-shutdowns-2021/.

5. 'Who is Nazanin Zaghari-Ratcliffe and why was she jailed in Iran?', *BBC,* 22 March 2022, available at: https://www.bbc.co.uk/news/uk-politics-42252741.

6. Dunleavy, P., 'UK's terror de-radicalization program continues to flounder', *the algemeiner,* 26 April 2022, available at: https://www.algemeiner.com/2022/04/26/uks-terror-de-radicalization-program-continues-to-flounder/.

7. Cobain, I., 'UK's Prevent counter-radicalisation policy "badly flawed"', *The Guardian,* 19 October 2016, available at: https://www.theguardian.com/uk-news/2016/oct/19/uks-prevent-counter-radicalisation-policy-badly-flawed.

8. Machon, A., *Spies, Lies and Whistleblowers: MI5, MI6 and the Shayler Affair,* Book Guild Publishing, 2005, Subversion chapter.

9. Lynskey, D., '"Alexa, are you invading my privacy?" – the dark side of our voice assistants', *The Guardian,* 9 October 2019, available at: https://www.theguardian.com/technology/2019/oct/09/alexa-are-you-invading-my-privacy-the-dark-side-of-our-voice-assistants.

10. St. John, A., 'Yes, your smart speaker is listening when it shouldn't', *Consumer Reports,* 9 July 2020, available at: https://www.consumerreports.org/smart-speakers/yes-your-smart-speaker-is-listening-when-it-should-not/.

Chapter 2

1. TechTarget contributor, 'Snowden effect', *TechTarget,* October 2015, available at: https://www.techtarget.com/whatis/definition/Snowden-effect.

2. Ackerman, S. and Roberts, D., 'NSA phone surveillance program likely unconstitutional, federal judge rules', *The Guardian,* 16 December 2013, available at: https://www.theguardian.com/world/2013/dec/16/nsa-phone-surveillance-likely-unconstitutional-judge.

3. Appelbaum, J., *To Protect And Infect Part 2: The militarization of the Internet,* media.ccc.de, 2014, available at: https://media.ccc.de/v/30C3_-_5713_-_en_-_saal_2_-_201312301130_-_to_protect_and_infect_part_2_-_jacob#t=2920.

4. Quinn, B., 'UK court approves extradition of Julian Assange to US', *The Guardian,* 20 April 2022, available at: https://www.theguardian.com/media/2022/apr/20/uk-court-approves-extradition-of-julian-assange-to-us.

5. Machon, A., 'Former MI6 spy v WikiLeaks editor: Who really deserves 1st Amendment protection?', *RT,* 24 August 2018,

available at: https://www.rt.com/op-ed/436761-christopher-steele-assange-first-amendment/.

6. *Collateral Murder*, WikiLeaks, 2010, available at: https://collateralmurder.wikileaks.org/.

Chapter 3

1. 'Biggest companies in the world 2022', FinanceCharts.com, accessed 4 May 2022, available at: https://www.financecharts.com/screener/biggest.

2. Aziz, S. 'Creeping counterterrorism: from Muslims to political protestors', *Truthout*, 22 August 2012, https://truthout.org/articles/creeping-counterterrorism-from-muslims-to-political-protesters/; 'City of London police occupy London domestic terrorism/extremism warning', Public Intelligence, 6 December 2011, https://publicintelligence.net/city-of-london-police-occupy-london-domestic-terrorismextremism-warning/.

3. Lewis, P., '"Our minds can be hijacked": the tech insiders who fear a smartphone dystopia', *The Guardian*, 6 October 2017, available at: https://www.theguardian.com/technology/2017/oct/05/smartphone-addiction-silicon-valley-dystopia.

4. Guadagnoli, G., 'The Netherlands is advancing the digitalisation of the government by turning open source theory into practice', *Joinup*, 15 April 2021, available at: https://joinup.ec.europa.eu/collection/open-source-observatory-osor/news/open-source-toolbox-released-netherlands.

Chapter 4

1. Malik, S., 'Occupy London's anger over police "terrorism" document', *The Guardian*, 5 December 2011, available at: https://www.theguardian.com/uk/2011/dec/05/occupy-london-police-terrorism-document.

2. Machon, A., 'One man's terrorist is another man's activist', Annie-Machon.ch, 6 February 2012, available at: https://anniemachon.ch/annie_machon/2012/02/one-mans-terrorist-is-another-mans-activist.html.

3. 'The NHS cyber attack', Acronis, 7 February 2020, available at: https://www.acronis.com/en-gb/blog/posts/nhs-cyber-attack/.

4. Kirchgaessner, S., 'Saudis behind NSO spyware attack on Jamal Khashoggi's family, leak suggests', The Guardian, 18 July 2021, available at: https://www.theguardian.com/world/2021/jul/18/nso-spyware-used-to-target-family-of-jamal-khashoggi-leaked-data-shows-saudis-pegasus.

5. Tonkin, S., 'Beware "the spy in your POCKET": How Pegasus software "found on a device connected to Number 10" lets hackers record calls, steal photos and secretly FILM you – and can be installed on any smartphone via a simple text message', Mail Online, 19 April 2022, available at: https://www.dailymail.co.uk/sciencetech/article-10731021/How-Pegasus-spyware-device-connected-Number-10-works.html.

Chapter 5

1. Messac, L., 'The other opioid crisis – people in poor countries can't get the pain medication they need', The Conversation, 25 March 2016, available at: https://theconversation.com/the-other-opioid-crisis-people-in-poor-countries-cant-get-the-pain-medication-they-need-56205

2. Shiva, V., 'The seeds of suicide: how Monsanto destroys farming', Global Research, 5 April 2013, available at: https://www.globalresearch.ca/the-seeds-of-suicide-how-monsanto-destroys-farming.

3. Gonggrijp, R. and Rieger, F., 'We lost the war: Welcome to the world of tomorrow', media.ccc.de, 2015, available at: https://media.ccc.de/v/22C3-920-en-we_lost_the_war.

4. Appelbaum, J., *To Protect And Infect Part 2: The militarization of the Internet,* media.ccc.de, 2014, available at: https://media.ccc .de/v/30C3_-_5713_-_en_-_saal_2_-_201312301130_-_to_protect_ and_infect_part_2_-_jacob#t=2920.

Chapter 6

1. Kochovski, A., 'Ransomware statistics, trends and facts for 2022 and beyond', *Cloudwards,* 22 March 2022, available at: https:// www.cloudwards.net/ransomware-statistics/.

2. Joseph, A., 'Are you buying your child a paedophile spy-cam for Christmas? How hackers are able to seize control of six top-selling gifts and tap into their video streams and microphones', *MailOnline,* 14 December 2017, available at: https://www.dailymail.co.uk/ news/article-5179201/Hackers-able-seize-control-childrens-Christmas-toys.html.

3. 'Toy firm VTech fined $650,000 over data breach', *BBC,* 9 January 2018, available at: https://www.bbc.co.uk/news/technology-42620717.

4. Ilevičius, P., 'Ring hacked: doorbell and camera security issues', *NordVPN,* 23 December 2021, available at: https://nordvpn.com/ blog/ring-doorbell-hack/.

5. Gallagher, R., 'How NSA spies abused their powers to snoop on girlfriends, lovers and first dates', *Slate,* 27 September 2013, available at: https://slate.com/technology/2013/09/loveint-how-nsa-spies-snooped-on-girlfriends-lovers-and-first-dates.html.

6. Cole, W., 'EncroChat: The shadowy Dutch "tech firm" that sold "encrypted phones" for "worry free communications" about "murder, buying kilos, guns and millions of pills" before disappearing without a trace', *MailOnline,* 2 July 2022, available at: https://www.dailymail.co.uk/news/article-8484673/EncroChat-shadowy-Dutch-tech-firm-sold-encrypted-phones.html.

Chapter 7

1. Simonite, T., 'A Zelensky deepfake was quickly defeated. The next one might not be', *Wired*, 17 March 2022, available at: https://www.wired.com/story/zelensky-deepfake-facebook-twitter-playbook/.

2. Machon, A., 'Libya: my enemy's enemy is my friend, until he becomes my enemy again . . .', AnnieMachon.ch, 24 March 2011, available at: https://anniemachon.ch/annie_machon/2011/03/libya-my-enemys-enemy-is-my-friend-until-he-becomes-my-enemy-again.html.

3. Campbell, D., 'British police are hounding a journalist for his sources – it's vital he resists', *The Guardian*, 18 February 2022, available at: https://www.theguardian.com/commentisfree/2022/feb/18/british-police-hounding-journalist-sources-chris-mullin-birmingham-six.

4. Campbell D., 'Birmingham pub bombings: Chris Mullins wins fight to protect source', *The Guardian*, 22 March 2022, available at: https://www.theguardian.com/media/2022/mar/22/birmingham-pub-bombings-chris-mullin-wins-fight-to-protect-source.

5. Parry, R., 'Pulling a J. Edgar Hoover on Trump', *Information Clearing House*, 12 January 2017, available at: http://www.informationclearinghouse.info/46214.htm.

6. Machon, A., 'Is the United States facing a coup d'etat?', *RT*, 18 December 2016, available at: https://www.rt.com/op-ed/370672-united-states-coup-trump-electoral-college/.

7. Chang, A., 'The Facebook and Cambridge Analytica scandal, explained with a simple diagram', *Vox*, 2 May 2018, available at: https://www.vox.com/policy-and-politics/2018/3/23/17151916/facebook-cambridge-analytica-trump-diagram.

8. Cadwalladr, C., '"I made Steve Bannon's psychological warfare tool": meet the data war whistleblower', *The Guardian*, 18 March 2018, available at: https://www.theguardian.com/news/2018/mar/17/data-war-whistleblower-christopher-wylie-faceook-nix-bannon-trump.

9. Cadwalladr, C. and Graham-Harrison, E., 'Revealed: 50 million Facebook profiles harvested for Cambridge Analytica in major data breach', *The Guardian*, 17 March 2018, available at: https://www.theguardian.com/news/2018/mar/17/cambridge-analytica-facebook-influence-us-election.

10. The WSWS Editorial Board, 'Google rigs searches to block access to World Socialist Web Site', World Socialist Web Site, 28 July 2017, available at: https://www.wsws.org/en/articles/2017/07/28/pers-j28.html.

11. Carlo, S., 'Big Brother Watch Briefing on facial recognition surveillance', *Big Brother Watch*, June 2020, pp. 15–16, available at: https://bigbrotherwatch.org.uk/wp-content/uploads/2020/06/Big-Brother-Watch-briefing-on-Facial-recognition-surveillance-June-2020.pdf.

12. Ibid., p. 16.

13. 'Young girl returned after kidnapping by man she met on Roblox', *BBC News*, 3 March 2022, available at: https://www.bbc.co.uk/news/world-us-canada-60607782.

Chapter 8

1. Greig, J., 'Viasat says "cyber event" is causing broadband outages across Europe', *ZDNet*, 28 February 2022, available at: https://www.zdnet.com/article/viasat-confirms-cyberattack-causing-outages-across-europe/.

2. Bannister, A., 'When the screens went black: How NotPetya taught Maersk to rely on resilience – not luck – to mitigate future cyber-attacks', *The Daily Swig*, 6 July 2019, available at: https://portswigger.net/daily-swig/when-the-screens-went-black-how-notpetya-taught-maersk-to-rely-on-resilience-not-luck-to-mitigate-future-cyber-attacks.

3. Collier, K., 'Starlink internet becomes a lifeline for Ukrainians', *NBC News*, 29 April 2022, available at: https://www.nbcnews.com/tech/security/elon-musks-starlink-internet-becomes-lifeline-ukrainians-rcna25360.

4. Buncombe, A., 'Chilcot report: Student whose thesis became Tony Blair's "dodgy dossier" accuses UK of systemic failure', *The Independent*, 6 July 2016, available at: https://www.independent.co.uk/news/uk/politics/chilcot-report-author-of-dodgy-dossier-accuses-uk-of-systematic-failure-a7123136.html.

5. Norton-Taylor, R., 'Iraq dossier drawn up to make case for war – intelligence officer', *The Guardian*, 12 May 2011, available at: https://www.theguardian.com/world/2011/may/12/iraq-dossier-case-for-war.

6. Sabbagh, D., 'Britain has offensive cyberwar capability, top general admits', *The Guardian*, 25 September 2020, available at: https://www.theguardian.com/technology/2020/sep/25/britain-has-offensive-cyberwar-capability-top-general-admits.

7. Voo, J., Hemani, I., Jones, S., DeSombre, W., Cassidy, D. and Schwarzenbach, A., 'National Cyber Power Index 2020', The Belfer Center, September 2020, available at: https://www.belfercenter.org/sites/default/files/2020-09/NCPI_2020.pdf.

Chapter 10

1. Patel, N., 'Hard lessons of the SolarWinds hack', *The Verge*, 26 January 2021, available at: https://www.theverge.com/

2021/1/26/22248631/solarwinds-hack-cybersecurity-us-menn-decoder-podcast.

2. Willett, M., 'Lessons of the SolarWinds hack', *IISS*, 31 March 2021, available at: https://www.iiss.org/blogs/survival-blog/2021/04/lessons-of-the-solarwinds-hack.

3. Nakashima, E. and Albergotti, R., 'The FBI wanted to unlock the San Bernardino shooter's iPhone. It turned to a little-known Australian firm', *The Washington Post*, 14 April 2021, available at: https://www.washingtonpost.com/technology/2021/04/14/azimuth-san-bernardino-apple-iphone-fbi/.

4. Johnson, R., '60 percent of small companies close within 6 months of being hacked', *Cybercrime Magazine*, 2 January 2019, available at: https://cybersecurityventures.com/60-percent-of-small-companies-close-within-6-months-of-being-hacked/.

5. Kochovski, A., 'Ransomware statistics, trends and facts for 2022 and beyond', *Cloudwards*, 22 March 2022, available at: https://www.cloudwards.net/ransomware-statistics/.

6. Gibbs, S., 'FBI doubts North Korea link to Sony Pictures hack', *The Guardian*, 10 December 2014, available at: https://www.theguardian.com/technology/2014/dec/10/fbi-doubts-north-korea-link-sony-pictures-hack.

7. Carnell Council CISSP, 'Can the voting process be hacked?', *Security Magazine*, 17 September 2020, available at: https://www.securitymagazine.com/articles/93385-can-the-voting-process-be-hacked.

8. Hern, A., 'Kids at hacking conference show how easily US elections could be sabotaged', *The Guardian*, 22 August 2018, available at: https://www.theguardian.com/technology/2018/aug/22/us-elections-hacking-voting-machines-def-con.

Chapter 11

1. Michon, A., 'LIBE whistleblower hearing at the European Parliament', Using Our Intelligence, 4 October 2013, available at: https://anniemachon.ch/annie_machon/2013/10/libe-whistleblower-hearing-at-the-european-parliament.html.

2. European Commission, 'Commission puts forward declaration on digital rights and principles for everyone in the EU', 26 January 2022, available at: https://ec.europa.eu/commission/presscorner/detail/en/IP_22_452.

ABOUT THE AUTHOR

Annie Machon is a former spy turned privacy campaigner. She was an intelligence officer for the UK's Security Service, MI5, before resigning to help blow the whistle on the crimes and incompetence of British spies in the 1990s.

She is now a media pundit, writer, and international public speaker on a wide variety of geopolitical issues, including the wars on terrorism, whistleblowers, drugs, the internet, press and media freedoms, secrecy legislation, civil liberties and accountability in government and business. Annie brings a rare perspective on not only the inner workings of governments, intelligence agencies and the media, but also the wider implications for the need for increased openness and accountability in both public and private sectors. She has also given evidence to both the European Parliament and the UK Parliament about surveillance in the wake of the Snowden disclosures.

In 2020 Annie was awarded the SA Award for Integrity in Intelligence by Sam Adams Associates, a global group of intelligence, diplomatic and military whistleblowers, served four years as the European Director of Law Enforcement Action Partnership, is a director of https://worldethicaldata.org/, an organiser of the World Ethical Data Forum, and is an advisory board member of the Courage Foundation.

She is also a regular contributor to films, most notably *Digital Dissidents*, *The Culture High*, *Disappear*, *The Mole*, and *Espionnes*.

Annie has an MA (Hons) Classics from Cambridge University and is the author of *Spies, Lies and Whistleblowers: MI5, MI6 and the Shayler Affair*.

INDEX

INDEX

INDEX

INDEX

INDEX

INDEX

INDEX

INDEX

INDEX

INDEX

INDEX

INDEX

INDEX